Shrinking the Cat

BOOKS BY SUE HUBBELL

FAR-FLUNG HUBBELL

A BOOK OF BEES

A COUNTRY YEAR

ON THIS HILLTOP

BROADSIDES FROM
THE OTHER ORDERS

WAITING FOR APHRODITE

SHRINKING THE CAT

Shrinking the Cat

Genetic Engineering Before
We Knew About Genes

SUE HUBBELL

With illustrations by Liddy Hubbell

HOUGHTON MIFFLIN COMPANY

BOSTON   NEW YORK

2001

For information about permission to reproduce selections from
this book, write to Permissions, Houghton Mifflin Company,
215 Park Avenue South, New York, New York 10003.

Visit our Web site: www.houghtonmifflinbooks.com.

Library of Congress Cataloging-in-Publication Data
Hubbell, Sue.
Shrinking the cat : genetic engineering before we knew about
genes / Sue Hubbell ; with illustrations by Liddy Hubbell.
p.   cm.
Includes bibliographical references and index.
ISBN 0-618-04027-7
1. Breeding. 2. Genetic engineering. I. Title
S494 .H83 2001
660'.65 — dc21   2001024547

Printed in the United States of America

Book design by Robert Overholtzer

QUM 10 9 8 7 6 5 4 3 2 1

The poem "What Would Make a Boy Think to Kill Bats," from *Dominion*
by Brooks Haxton, copyright © 1986 by Brooks Haxton, is reprinted by
permission of Alfred A. Knopf, a Division of Random House Inc.

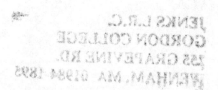

For my old friend ARLEN EDGAR, who
observed so many years ago that *Phalangium opilio*
was man's closest phalangid fellow traveler.
This book has been brewing
ever since.

CONTENTS

A map of the Silk Roads is on pages 60–61.

Creation is not an act but a process; it did not happen five or six thousand years ago but is going on before our eyes.

— Theodosius Dobzhansky, *Changing Man*, 1968

PREFACE

Genetics. A branch of biology that deals with the heredity
and variation of organisms and with the mechanisms
by which these are effected.

Engineering. The science by which the properties of
matter . . . in nature are made useful to man . . .

— *Webster's Third New International Dictionary*

I'd not intended to write another book for publication after I
finished my last one, *Waiting for Aphrodite.* That book had
seemed final to me, and I wanted to devote myself to other
things — working in my woodlot, for instance, or building
stone walkways around the house.

But as I worked, thinning out trees here and there, encouraging those I liked, cutting down those I didn't, hauling rock
from this place to that, I reflected on my proclivity for re-
arranging bits of the world, an activity so characteristic of the
human animal. It is a subject I've written about in the past
but clearly have not exhausted.

About that time, stories concerning genetically modified

organisms began to appear in newspapers everywhere. I read them with interest, not only because they were good examples of our human penchant for fiddling with materials that are to hand but because all parties to the furor that erupted seemed to be talking past one another.

The public, which had forgotten whatever high school biology it had learned, was saying that something new, terrifying, and possibly devious had been kept from it, something having to do with the sanctity of species. The scientists, whose work was built on a body of genetics research stretching back a hundred years, appeared startled to find that the term "species" was understood by the public in such a fixed way. Biologists, for whom "species" had become simply a useful word, were used to reassigning plants and animals to different species. They knew that the genetic similarities among species were far more important than their differences. They saw the uniformities of biological processes as transcending the separateness of individuals. Biologists had their own questions about genetic engineering, questions that weren't making it into the popular media, but they understood that manipulating a gene or even putting a gene from one kind of life into another wasn't such a stretch. They had been saying such things for some time, but in words so obscure and papers so technical that no one outside their particular fields had heard them.

The corporations that were exploiting genetic research, had begun profiting from it, and had every expectation of profiting even more were alarmed by all the attention and turned, as structured organizations always do, to spin control, which fooled no one and made the public doubly suspicious.

I knew, perhaps, a little more than the general public did about genetic engineering, but certainly not as much as the scientists. I knew, for instance, that we had been fiddling with the genetic identities of domesticated plants and animals ever since we had become human. I knew that in the process of that fiddling — engineering by another name — we had actually created brand-new kinds of life, species, if you will, based on but different from the wild forms that had furnished the raw material — wheat and corn, for instance, to name but two. And I knew that to a greater or lesser degree those new species needed us in order to thrive. The enormous sums of money being invested in research relating to modern genetic engineering by agribusiness and pharmaceutical corporations, sums they believed would be repaid many times, seemed like a new element in the story. But in truth, merchants and traders from ancient times onward had been in the business of bringing more desirable, and hence more profitable, goods from there to here: amphoras of better olive oil brought higher prices in places where lesser olives grew. Sheep with denser woolly coats could be sold to advantage far from their birthplace.

But I had a lot of questions. How had merchants spread the new plants and animals from one end of the earth to the other? What was the effect of transplantation on the plants and animals? Could those species whose unique genetic makeup was the result of our handiwork live without us? How did all those genetically modified organisms affect human history? Until recently, the genetic changes that created new species had been brought about through breeding, through the isolation and encouragement of genetically interesting recessive genes and mutant plants and animals, and through the artificial creation of mutations themselves. From the standpoint of natural selection, artificial selection speeded up the process of evolutionary change. Still, species change had been slow compared to what researchers today can do, with our knowledge of genetic processes and the tools we have to manipulate those processes directly. Does the modification of species in the past have anything instructive to tell us about the moral and ethical questions concerning modern genetic engineering?

Those questions interested me, so for a time I put off work in my woodlot and let the rocks lie. To answer my questions, I chose a few animals and plants whose genetic identities we have tinkered with to varying degrees and that are more or less dependent upon humans. In the course of examining our shared histories, I discovered that our experiments in manipulating species have had unintended consequences. I've answered my original questions to some degree, but I've found a lot of new ones. These relate to an enormous problem confronting humankind today. For the purposes of this book I will call it the problem of limits: How do we limit the effects of six billion of our kind on the rest of the world and avoid making alterations that harm other kinds of life and change the world so drastically that we can no longer live in it ourselves?

I am encouraged and hopeful about this large, general, and occasionally boisterous public debate over genetic engineering. I believe it is long overdue, and I also believe that it will become, in time, a way to and a part of the solution of the problem of limits.

ACKNOWLEDGMENTS

One of the pleasures of working on this book was that it gave me an excuse to use some fine libraries with which I had not been acquainted. I owe them and their staffs a debt that I should like to acknowledge here: the Jackson Laboratory library in Bar Harbor, Maine; the National Agricultural Library; and the National Library of Medicine. Two librarians at familiar libraries helped in special ways. Morna Bell, a magician with interlibrary loans at my hometown library in Maine, and Diana Niskern, a science specialist at the Library of Congress. Oftentimes Diana appeared at my favorite desk staggering under a load of materials, saying, "I know we found what you were looking for yesterday, but I thought you might find some of this interesting, too." Indeed, I often did. Life can't be all bad when such a beautiful place as the Library of Congress is dedicated to the preservation of knowledge and has people like Diana working there.

Others helped at specific points, and I thank them. Christine Sarbanes, a fine classicist, helped me name humankind. Howard Mansfield's research on Johnny Appleseed laid a foundation for my own. Dilys Winograd helped me get in touch with an elusive source, as did Marion Stocking. Helen Ghiradella and Edwin Barkdoll made good recommendations about scientific sources and had the generosity to read through a draft of the text and make helpful suggestions. Those whom I interviewed gave generously of their time and also took time to review their contributions. The mistakes that remain, however, are of my own doing.

Harry Foster, editor and neighbor, has performed his usual brilliant job in the birth of the book. Harry, where have you been all my life? I am grateful to you, as I am to the other half of the Houghton Mifflin editing team, Peg Anderson, text editor supreme, who has the extraordinary ability to see both forest and trees at the same time.

Liddy Hubbell's drawings enhance the text, and Brian Hubbell's index is an unusually satisfying one. Thank you both.

And to Arne ("Can I get you anything today? Can I do something for you?") Sieverts: your supple photocopying has no equal, and I thank you for always being there.

Shrinking the Cat

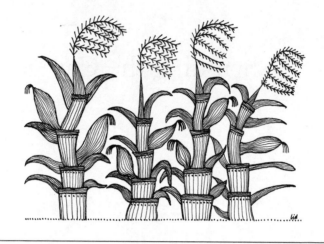

[O N E]

Of Humanity,
Tazzie the Good Dog,
and Corn

"I make thee maister," seid Robyn Hode.
"Nay . . . let me be a felow," seid Litull John.
— *Robyn Hode and Monk*, 1450

W E, THE NAMERS, call our species *Homo sapiens,* the sapient, intelligent, wise sort of human. It is the name by which we distinguish ourselves from all other kinds of life, including those other species of the genus *Homo: H. erectus,* those who stood up, *H. habilis,* those who made stuff. They, along with the australopithecines, Cro-Magnons, and other relatives of our fine selves, managed to get themselves extinguished somewhere along the line, but we thrived and continue to do so. How wise.

Tens of thousands of years of natural selection acted upon our ancestors to produce our species of humankind with our complicated, much-folded brains and clever fingers, attri-

---

OPPOSITE: Peruvian corn. Adapted from a drawing of a decorative frieze on pottery, c. 2000 B.C., in Walter Lehman, ed., *Art of Old Peru.*

butes that have allowed us to spread all across the planet and even a little beyond it. In biological terms, we have a very big ecological niche.

We've all seen those lists that appear in the newspapers now and again ranking animals by intelligence, giving the pig a higher grade than the horse, and the pigeon higher than either. On those lists, of course, we always come out best of all, outranking even the animals we regard as suspiciously bright, such as dolphins or apes. We win because we make the rules: we define intelligence as our kind of knowingness and dismiss any other animal knowingness as mechanical, instinctual, "hard-wired." That is the sort of verbal sleight of hand I would have been graded down for in high school debating competitions because it is an error of circularity. We define intelligence by holding up other animal minds to ours and to our way of understanding, then dismiss their ways by giving them a lesser name.

Considering all the world's other intellectual follies, the pop biology represented by these lists hardly deserves to be fussed about. Nevertheless, it is heartening to note that there have always been dissenters from the conceited notion that we are the wisest animals on the planet.

Margaret Cavendish, a seventeenth-century Englishwoman, a bluestocking condemned by Samuel Pepys as "ridiculous" and praised by Charles Lamb as "princely," wrote:

> For what man knows whether fish do not know more of the nature of water, and ebbing and flowing and the saltiness of the sea? Or whether birds do not know more of the nature and degrees of air, or the cause of tempests? Or whether worms do not

know more of the nature of earth and how plants are produced? Or bees of the several sorts of juices of flower than man? . . . Man may have one way of knowledge . . . and other creatures another way, and yet other creatures' manner of way may be as intelligible and instructive to each other as Man's.

As a human, I've never been flattered by my own exaltation in those intelligence-ratings lists, for I've spent my life with other animals and have come to know that they have considerable capabilities that I lack. I don't expect them to take human IQ tests, because they don't live human lives. They live dog, cat, honeybee, monkey, lion, cricket, salmon, squirrel, pigeon lives, and I'd fail their IQ tests as readily as they would fail mine. For instance, I'm no good at bones.

Some years ago I shared my life with a dog named Tazzie the Good, a Belgian shepherd–black lab cross. Tazzie had all the genetically determined traits that dog breeders had drawn out over generations of inbreeding her ancestors, along with some hybridism that juggled those traits a little. More thoughtful than the average lab and a lot mellower than the average shepherd, she was obedient and devotedly loyal to me, wanting only to learn what I wanted from her and then joyfully doing it. She accepted other humans with grace and dignity. She was the definition of Good Dog. She had massive jaws that could have killed smaller creatures easily, but she used them gently because I required it. When Black Edith Kitty was a mere kitten, he would tease Tazzie into playing with him. (Like the boy in the Johnny Cash song, Black Edith has a girl's name.) Tazzie would hold him between her paws as though the cat were a living bone and delicately gnaw on him. When Black Edith had had enough of this slobbery sport, he would use his claws to rake Tazzie across her soft black nose, and she would let him go.

Even compared to other dogs I have known over the years, Tazzie had extremely high bone intelligence. My husband and I have a house in a tidy neighborhood in Washington, D.C. No one there is a trashy sort of person who would strew garbage about or leave it in an uncovered bin. Yet when we took her on evening walks around the block, Tazzie would *always* find a bone. Not one night, not occasionally, but every night. If I'd been sent out on a treasure hunt and told that my mission was to find, in the dark, a bone in a one-block circuit of that neighborhood, I'd have failed — not just once but night after night.

Tazzie was bright about bones. I am not. Tazzie was sapient in a dog's world, many aspects of which are not mine. Her world and mine overlapped at certain points — we were companions — but our skills, our intelligences, were for our own ways of life, not each other's. And those ways of life were determined by different biologies, different configurations of our DNA.

Knowingness — call it intelligence, call it instinct, call it whatever you will — is always knowingness for some purpose. It is not an abstraction, not some objective gold standard. I'd never have expected Tazzie to discuss the finer points of the role of NATO in the post–Cold War world, and she accepted that I was stupid about bones, although she never quite gave up trying to interest me in the stone game. That was a game she had invented as a puppy, when we lived beside a river in the Missouri Ozarks. She would select a stone and drop it in my lap to throw for her. I'd toss it into the river, and she would leap through the air following the stone's trajectory and watch it splash and disappear. She had no interest in retrieving it. She was not that kind of a dog. Instead she'd bring a *new* stone to me and whine enticingly, urging me to throw it. As she aged and grew stiffer, she had as much fun simply sitting beside me and watching me throw stones. Her eyes would follow the stone's arc and its splash, and then she would whine and beg me to throw another. I always got bored with the stone game before she did, but now that she is gone I wish I'd spent more time chunking stones into the water for her and less time talking about NATO at Washington dinner parties.

Some time has passed since Tazzie died in the fullness of age, but I still miss her. She was my shadow. I can still feel her chin on my knee when I am driving the car, her quiet paw on my foot as I sit writing. She was a very *present* animal.

All the years I've spent hanging out with other animals and the zoologists who study them have led me to think that it is arrogant and even silly of us to name ourselves for our intelligence. But we have another preeminent trait, one that we are good at beyond all other animals, and that is our ability to modify the world to make it nice for ourselves. A more accurate name for us might be *Homo mutans,* the human who changes things. We do not accept what we find. We use that brain and those fingers, with the extensions we've made for them, to alter, build, excavate, extend, recombine, fuss, and fiddle to make the world comfortable and more interesting to us.

Other animals modify their particular worlds, of course, to

the extent that they are able. Honeybees harvest plant sap and process it into a glue to chink their hives against winter drafts. Bowerbirds gather up fresh flowers to display in the competition for mates. Caddis fly larvae make protective shelters out of grains of sand. Last winter a mouse stole all the little wooden vegetables from a dollhouse I have for grandchildren. I do not understand the mouse's purpose: when I pulled open one of my desk drawers I found within it a mouse nest made of pulled chair stuffing surrounded by the tiny wooden carrots, cabbages, and rutabagas. Both chimpanzees and certain insect-eating finches use twigs to stir up ants and other bugs from their nests so that they are easier to catch and eat. I remember a bold tufted titmouse who made life uncomfortable for a friend of mine by making unrelenting, swooping passes over his curly head of hair and then, with great daring, pulling out several hairs. We watched as it skillfully wove them into a nest.

Tinkering with existing circumstances and materials is characteristic of life in all its forms, but we humans have better biological equipment for this purpose. And as a result we are more thoroughgoing. We are the fiddlingest animal the world has ever seen.

Now that we are six billion in number and have more powerful tools than ever to modify the planet, we are starting to wonder whether we are fiddling so much that we and other kinds of life may not be able to continue living on it. We are finally beginning to ask where all this fiddling is taking us. Impatient environmentalists sometimes get discouraged and think we aren't getting anywhere, but we've only recently become aware of the problem and have just commenced asking questions. It is only about one hundred years since we filled up the habitable earth and far fewer since we realized that there are too many of us and only just now that we are attending to the fact that we do not have an entirely happy effect on other kinds of life.

We will have to ask a lot of questions before we can figure out how to frame those that might help us solve the problems we have caused. And we are just beginning to pose the necessary questions about limits: limits to our own numbers, limits to the extension of human life into prolonged senescence, limits to our consumption and exploitation. These are ferociously difficult questions because our biological design is to increase willy-nilly and to grab whatever is available (as it is for any species). But we are developing a modicum of self-awareness and a struggling, if not always successful, objectivity, and that makes me optimistic about our kind. It may be that none of us alive today will be around long enough to see solutions to the problems of limits, but I think our descendants will find them. And when they do, we will deserve to be called *sapiens*.

One of the questions we've asked as we began to consider our impact on the world runs something like this: Weren't there — somewhere, sometime — pure and unspoiled people who made better arrangements with the world than we have? Aren't today's "primitive" people rather like that? The answer to both questions is, probably not. As Jared Diamond, the physiologist who has studied and thought about people in out-of-the-way places, often points out, people everywhere are as exploitative of the resources they find as their technology and numbers permit. It is easy to believe that those who are fewer in number somehow have better intentions toward the planet than we.

That river where I used to chuck stones for Tazzie is named the Jack's Fork. Part of the first National Scenic Riverway designated by the U.S. Park Service, it runs through rugged land in the Ozarks of southern Missouri. While I lived there I came to know a federal archaeologist who had excavated along its banks and had found that a long time ago — I no longer remember the dates — there had been densely populated Indian villages along the river. The Indians had cleared the land and farmed the thin soil. Farming in the Ozarks is a heartbreaking business, as the white settlers were to discover. The land is too steep and the soil too thin and poor for good crops. The archaeologist had found artifacts — pots, bowls, baskets, and arrow points — in the remains of the Indian houses. The oldest artifacts were made with expertness and grace, but the most recent ones lacked both qualities. He had arranged a collection of arrow points by date of making, and they became cruder and cruder over time; the baskets and pots also became increasingly rougher and coarser, as if they had been made by a people who had grown sad and ineffectual. There were signs that in the end the people had left quickly. The archaeologist suspected that disease of some kind had swept through the towns (as evidence suggests happened elsewhere in North America) and that those left were too few or too weak to defend themselves.

Dust and dead leaves would have blown through the empty villages (some of which had been burned). Gradually over the years, their garbage dumps and home debris, their tools and pots, were covered and buried under accumulating soil. Weeds crept into the fields where they had grown corn, a crop Europeans had yet to learn of. Low bushes, then trees followed. The Jack's Fork, no longer silty from farmland run-off, ran clear. Slowly the Ozarks took on the character it had when Americans of European descent found it in the 1800s, a wild place full of big pine, oak, walnut, cedar, and hickory trees, clear streams, and wild animals. There were no signs left to tell that it had once been home to many people who had used everything they found there.

The story was repeated in many parts of the Americas. Peoples from the Mayans, with their high culture, to the Mound Builders increased in numbers, altering and using whatever they could, until life became precarious for one reason or another and their ways failed them. Their descendants, few in number, lived on in poorer circumstances.

The Indians whom the Europeans first met when they came to North America, even if they were descendants of people who had known greater amenities, had more skill at living from the land than the Europeans, who were mainly city folk with no knowledge of fishing, hunting, or farming. The native plants — wild berries, nuts, pawpaws, and the like — were not enough to sustain them, and they did not even recognize some of what was available as food. Lobsters, for instance, they regarded as offal. In 1622 William Bradford, one of the early Pilgrim dissidents who had come to these shores on the *Mayflower,* wrote a letter describing the settlers' privations, using as an example their humiliation and sadness when, on the occasion of receiving visitors, "the sole dish they could present their friends with was lobster."

Some settlers starved, but some took lessons from the Indians, who were farming several staples — sweet potatoes, squash, and corn — unknown to the Europeans. They quickly learned how to grow them. None of these crops was native to North America; they had all come from the New World tropics and had been cultivated and improved over the centuries. Of the three, corn would become the most important foodstuff that the New World gave to the Old.

Corn, like squash and sweet potatoes, was the result of the sort of fiddling with materials at hand that has taken place all over the world and is now called agriculture: the human creation of new botanical species whose genetic structures are distinct from those of their wild ancestors.

Corn is basically a grass, a member of a large family of plants that includes not only the green kind growing as lawns but also bamboo, rice, and sugar cane. Corn betrays its origins as a neotropical grass by a peculiarity of leaf anatomy — the clustering of photosynthetic cells, the engines that turn sunlight into plant energy — around the leaf's midrib. This configuration is typical of plants growing in hot, dry places. In temperate-zone plants, by contrast, the photosynthetic cells are ranged around the leaf veins, where they can respire in cooler temperatures.

The earliest farmers selected certain plants, some of which were mutants, that produced unusually good things to eat and saved their seeds to plant again. They planted the selected seeds, cosseted the plants that grew from them, and repeated the process again and again until they had created crops that were dependable and productive. Those traits are genetically determined, and thus those early farmers were re-arranging genes, even though they did not know what genes were.

The ancestor of corn is lost in some pre-Incan, pre-Mayan, pre-Aztec past, but a respectable hypothesis that accounts for many aspects of the corn plant has been put forward by Hugh H. Iltis, a botanist at the University of Wisconsin, who has spent much of his research life thinking about the history of corn. He conjectures that corn originated 7,500 years ago in Central America as a botanical anomaly when an exist-ing grass (probably quite like corn's nearest modern relative, called teosinte) had its male parts hijacked by its female parts. As a result, the plant produced very large seeds growing against a protective sheath (which would in time become a husk). They looked more like tiny kernels of corn than like ordinary grass seeds. Such a transformation may sound bi-zarre, and it is uncommon, but it is an expression of relic bi-sexuality that sometimes does take place in plants during pe-riods of stress — during an unusually wet rainy season, for instance. And that sort of stress is just what may have started corn on its way, according to Iltis. He adds that in the wild, when the weather pattern returns to normal, this relic bisex-uality would disappear.

Strictly speaking, these proto-corn plants were, as Iltis phrases it, a genetic "catastrophe," hoarders of female energy in the form of seed. But Indian farmers kept the seeds and planted them in rows, which were more convenient to tend and weed than scattered plants. They must have continued to

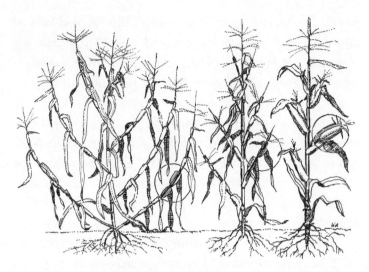

The possible transformation of corn in three steps to an upright-growing plant. Circled areas show enlargement of female parts. Adapted from Hugh H. Iltis, "From Teosinte to Maize."

select and plant seeds from plants that grew straighter and sprawled less than the original grasslike forms, because the plants genetically inclined to slim tallness were easier to farm. In the course of time this wonderful new crop plant was carried northward. The earliest archaeological evidence of corn in North America is about two or three thousand years old. Tiny kernels of popcorn were found in a cave in New Mexico along with tools and other signs that a family had lived there. The kernels, the discoverers found, could still be popped.

Those innovative early farmers would have been quickly encouraged in their agricultural efforts because the food value of corn rapidly improves with selective planting, as was demonstrated in an experiment conducted at the University of Illinois. The agronomists selected and planted corn seeds that were slightly higher in oil content — a genetically determined trait — than the rest of the seeds. They repeated the process after each harvest, always saving the seed with the highest oil content for replanting. When they graphed the results, they found a steady upward curve that showed no sign of leveling off when they brought the experiment to a close after fifty generations.

Even though those early farmers couldn't reach into the grass-becoming-a-corn-plant and fiddle with its genes as we can today, they could see and taste the expression of its genes. Biologists call the visible package of expressed genes the phenotype. When you look at me or I look at you, we see a phenotype. The stalk of corn, with its coarse, bladelike leaves, its silk, its corn kernels neatly and greenly wrapped, is the phenotype of its expressed genes. Not all of its genes are expressed, nor are all of yours or mine. All of us — humans, Tazzie the Good Dog, a corn plant — carry a unique genetic profile in our cells. That profile includes both expressed genes and ones that are not expressed: genes that are nonfunctional (as far as we can tell) and those that are recessive and thus masked by the dominant form of the same gene. All of those genes — unfunctional, expressed, and unexpressed, dominant and recessive — make up an individual's genotype. Genotype. Phenotype. It oversimplifies, but it helps to think of a package delivered by the UPS man. The package is the phenotype. You can tell something about what is inside by the size of the box, its weight, and the way it does or does not rattle. But you can't tell exactly what is inside until you open it. The contents are the genotype.

The individual genes for any plant or animal can take different alternative forms, which are called alleles. In sexual pairing, genetic material from both parents is shared to make up a new individual. In humans, for instance, brown eye color is controlled by a dominant allele and blue eye color by a recessive one. Two brown-eyed parents may each carry a blue eye-color allele, and if, in the genetic sorting that takes place in mating, their baby receives both blue-eyed alleles, she will have blue eyes. Their baby has a phenotype all her own.

It was only about one hundred and fifty years ago that the Austrian cleric Gregor Mendel began to elucidate this process, and it is startling to realize that the word "gene," so much a part of contemporary vocabulary, was coined in its modern sense only in 1909 (by Wilhelm Johannsen, a Danish agronomist).

Inheritance, and hence genetics, can be complicated for a number of biological reasons, and when Mendel began his work on what he called "factors of inheritance," he had the bad luck to choose as his first experimental model an animal that featured some of those complications: the honeybee. His experiments ended with confusing results, but fortunately he was a persistent man. Drawing on his interest in plant hybridizing, Mendel chose peas as his next experimental model. And peas, although Mendel could not have known it beforehand, are, fortunately, simpler genetically. With his pea experiments he demonstrated the emergence of recessive characteristics. Peas with red flowers and long stems carry dominant alleles of certain genes, and when they are crossed with white-flowered, short-stemmed peas, which have recessive alleles of the same genes, the first generation of plants all has red flowers and long stems. But when that first generation is intermated, the white flowers and short stems reappear in the next generation. And, what is more, they reappear in a tidy, predictable ratio. (Some suspect that his results were a little too good, that he fudged a little.) Later Mendel tried to test his conclusions on other plants and, once again, was left with more questions than answers and grew discouraged.

A basic understanding of how recessive traits are expressed, along with a new understanding of some aspects of inheritance that bewildered Mendel, has become the stock in trade of horticulturists, animal breeders, and agronomists.

In the wild, natural selection nudges and prunes the genotypes of plants and animals toward toughness, vigor, and successful reproduction and in that process maintains within the population a characteristic ratio of allelic frequency. But when artificial selection takes over, the story changes. One of the tricks that breeders of plants and animals had learned long before Mendel was that close inbreeding, although it might decrease vigor, could bring out traits we liked. Close matings to bring out recessive alleles and preserve mutations have created dogs with floppy ears, cats with no hair, and pigs with extra loin-end vertebrae and hence extra pork chops. Those attributes are ones that we fancy, even if the wild, harsh world does not.

Wild grass, corn's ancestor, shed its seeds abundantly and freely. Biologists think of corn as a "botanical monster," for its seeds, wrapped in unnaturally tight husks, do not disperse and plant themselves. And, even if a husked ear falls and is buried in the soil, the young shoots growing from the kernels die because they are overcrowded. But those tightly clustered kernels in their waterproof husks are just what we want. Never mind that corn does not replant itself. We will plant it, just as we will take care of its lack of vigor, its vulnerability to pests and diseases. Corn is a man-made plant, and in return for the good features we have drawn from its genotype, we will take care of it.

If little green men were to swoop down from the sky one day and kidnap all of humanity in their spaceships, our descendants — brought back to the planet after five thousand years of good behavior — would find no corn. Corn has to have human beings if it is to live and grow and reproduce.

After those first grateful European settlers learned how to grow corn, it soon became a sustaining crop. They could feed it to livestock or eat it directly. They could grind it into meal, which could be boiled into mush or fried into journey cake. A mash prepared from corn made an Everyman's tipple. Over the several hundred years since we began to grow it, new varieties of corn have been developed to suit new purposes. As anyone who drives through the Midwest in summertime might suspect, corn is the number-one grain grown in the United States. Just under ten billion bushels of it are grown every year — three times as much as its nearest rival, soybeans. Most of this harvest isn't the sweet corn sold at Mr. and Mrs. Smith's roadside stand. It is corn for animal feed, corn for ethanol, corn for starch, and corn for syrups to feed the world's seemingly insatiable taste for sweet snack foods and drinks.

That summertime driver can't help but notice how much growing room corn takes up: the fields look endless. The United States is the world's biggest producer of corn and devotes more acreage to it — more than seventy-two million acres — than any other country does. But extensive stands of corn (or any other plant) are a wonderfully attractive lure for pesty creatures and diseases, which sustain themselves and multiply on its bounty.

Corn borers are corn growers' most worrisome insect, hard to treat because they penetrate the plant and do not stay long on the outside, where they can be conveniently sprayed to death. They are the caterpillars of a modest-looking gray-brown moth, *Ostrinia nubilalis,* and they tunnel through the cornstalks and prop roots, even infesting the tassels of the ears. The moth came from Europe, where it dines on other plants as well and is less of a pest, not only because less corn is grown there but also because it has been there for a very long time and is kept in check by a variety of diseases and predators with which it has evolved. In North America the caterpillars found corn much to their liking and thrived in the absence of predators.

Over the years, agronomists have found a number of ways to deal with corn borers. Birds will eat the moths' eggs as well as the caterpillars, and when corn is grown in small plots with hedgerows around them, birds can find places to nest nearby. A variety of insecticidal sprays, including some that are illegal in the United States today, such as DDT, were used with modest success. Some cultural practices, such as plowing and disposing of infested stalks, helped when farms had enough hands to do those chores. Certain strains of corn show more resistance to borers than others, and for a while agronomists worked at developing that resistance genetically. They also imported, from Europe, disease organisms and predators that attack corn borers with fair success. All of those practices are more helpful for small stands of corn than for large ones. But small plantings are more labor-intensive than current agribusiness profit margins allow.

Our talent for fiddling with materials to hand — making up corn in the first place, concentrating it in monoculture stands, and inadvertently bringing corn borers to it — created a problem. *Homo mutans* is good at finding a solution to a specific, limited problem by bringing in another technological fix. Up until half a century ago, we solved problems of this sort genetically by crossbreeding — some of it pretty fancy, to be sure — and by manipulating the reproductive process to bring out the expression of genes in which we were interested. We may have had a more sophisticated understanding of the genetic determination of characteristics than the Indian farmers who developed corn those thousands of years ago, but essentially we were continuing to do what they had done: taking stock of the phenotype.

But in the past fifty years, as the structure of DNA has been elucidated and molecular biologists have learned what happens when gene switches are flipped, biotechnicians have learned how to reach inside cells and fiddle directly with those processes. The fields of study that have grown from this new understanding are exciting and use a lot of shiny, expensive machinery. They have drawn many of our best young biologists, because it is easy to get funding for their work from the agribusiness and pharmaceutical corporations that stand to gain from their discoveries.

What we have learned to call recombinant DNA technology (which is, essentially, reaching inside and fiddling with the genes directly) has begun to produce profitable substances, including hormones such as insulin for diabetics. That invention was helpful for humans and lucrative for the corporations. The agricultural biotech companies used the same techniques to produce bovine growth hormone. BGH obviously provides financial benefits not only to its manufacturers but to industrial dairies (more milk from fewer cows), despite the physiological price paid by the particular bovines into whom it is injected.

Biotechnology in the past several decades has been able to slip interesting genes from one organism into the genotype of another. In crop plants and animals, this produces changes much more precisely, directly, and quickly than does crossbreeding. And this form of fiddling we have learned to call transgenic engineering. A bacterium, *Agrobacterium tumefaciens* (which in its original host causes the disease crown gall), was used to shoehorn into the genotype of tobacco plants certain genes that would provide resistance to diseases, insects, and herbicides. That worked so nicely that the same method has been used with other crop plants. Biotechnicians learned to put extra growth genes into fish farmed in captivity to make them grow faster and turn a quicker profit. They put a moth gene into apples to produce a degree of resistance to fire blight. Monsanto, one of the agribusiness giants, is developing a corn plant that may produce plastics, even pharmaceuticals, courtesy of a cluster of bacterial genes. Using a similar method, the seed company Asgrow has developed squash seeds that are resistant to disease. By 1999 half of the soybeans growing in U.S. fields had been genetically rejiggered to withstand weedkillers.

Until recently, all of these developments sounded like good news to many people, yet another happy result of sparkling new technologies. To agronomists, developing plants with a genetic resistance to weedkillers seemed a particularly forward-thinking thing to do. A couple of generations ago, plowing and cultivating kept the weeds in farmers' fields in check. But those techniques kill earthworms, which help increase the soil's fertility. Plowing and cultivating also make the soil fluffy and loose so that the wind can blow it away, as it did in the Dust Bowl drought in the 1930s. In the aftermath of that disaster, agronomists invented the way most agriculture is practiced today, the no-till method. With the plow and the tiller on the scrap heap, earthworms thrive, the soil stays in place, and less labor is required to grow crops. But when soil is not plowed and tilled, weeds proliferate, choking out crop plants. As a result, weedkillers — herbicides — have replaced cultivation. But weedkillers are not selective, and unless care is taken, they can kill crop plants, too. The agronomists reasoned that it would be useful to put weedkiller resistance directly into the plant genes so that herbicides could be applied widely, saving on the labor costs of repeated precision dousings.

The companies developing this method weren't trying to keep it secret, but the public showed very little interest in what seemed to be rather boring laboratory matters. But then agriculture researchers at Monsanto took a gene derived from a bacterium and fitted it into corn. The bacterium, *Bacillus thuringiensis,* usually called *Bt* for short, produces a moth-killing toxin, and should thus, the researchers reasoned, confer upon corn the ability to produce its own pesticide against borers. This would be profitable to farmers because they would not have to apply expensive chemical pesticides. It would save on labor costs, too, and the lessened dependence on chemicals could be presented as an environmentally friendly development.

Commercial insecticidal sprays and powders made from *Bt* have been around since the 1960s. They were invented in part because of the public reaction to Rachel Carson's book *Silent Spring,* which had created a strong aversion to chemical pest controls and an interest in biological controls, which were thought to be more benign. *Bt,* as a naturally occurring bacterium, became over the years one of the most popular treatments against gypsy moth larvae and other unpopular caterpillars. Even organic gardeners found it acceptable. But entomologists have grumbled about *Bt* for years, for they knew that it was toxic not only to the caterpillars we call bad but to those of a great number of other moths and butterflies, too. In addition, entomologists kept collecting evidence showing that widespread use of *Bt* was encouraging the reproduction of several species of moths whose caterpillars were resistant to it. Resistance is a genetic trait, and *Bt* on its own was serving as a genetic modifier in the old traditional way of natural selection. *Bt* didn't kill *all* the caterpillars, just those that didn't have the genetic makeup to resist it. Those that were resistant multiplied even faster.

Nevertheless, Monsanto developed corn laced with *Bt* genes, and the crops planted from the seeds did indeed, at least temporarily, kill off many corn borers.

Monsanto was happy. It sold a lot of seed. The growers were happy. Fields planted with the seed showed a lower density of corn borer infestation. But then a piece of research done by a couple of entomologists at Cornell University was picked up by the press. The researchers, carefully choosing one of the few insects that Americans like, the orange and black monarch butterfly, had fed monarch caterpillars milkweed leaves (their preferred food) that had been dusted with pollen from the *Bt*-corn plants. Under laboratory conditions, half the experimental group of caterpillars died. Laboratory conditions are different from field conditions in that the experimental caterpillars had no choice of what they could eat. Entomologists were quick to point out the flaws in the experiment, but they knew that in general, *Bt* was bad stuff for caterpillars of all kinds.

The media did not have the time, space, or understanding for nuance, and the story quickly played as FRANKENFOOD KILLS BUTTERFLIES!!!! A suspicious public in the United States seized on the story, joining Europeans who were already nervous about genetically altered food. Clearly the consequences of genetic engineering, of rearranging parts of life that were too small to see and mysterious in their workings, were scary to many people.

Protesters dressed up in butterfly costumes. In Maine activists used machetes to cut down a field of experimental corn in which researchers were testing genetically induced weedkiller resistance. Industrial food processors, fearing boycotts and loss of revenue, vowed not to use genetically engineered foodstuffs.

It is understandable that those of us who are not biologists have a shuddery reaction to stories about anything that is transgenetically altered. It seems unnatural because we are accustomed to thinking of species as categories of reality: a bacterium is one thing and corn quite another; cows and tobacco and fish are entirely discrete bundles of life, all politely separated from one another . . . and from us, for goodness sake! If someone in a white lab coat who uses a lot of Latinate words can move a gene — something we don't quite understand but that we know is important in making each kind of life special — from one species to another, well, that makes us feel penetrable and unbounded and a little queasy. FRANKENFOOD KILLS BUTTERFLIES!!! What about us?

That reaction and that question are not surprising, considering that many people, including policymakers, don't really understand what molecular biologists have been up to in recent years. Part of the reason for that lack of understanding is that the biologists haven't done a very good job of explaining their research in terms that nonscientists can comprehend.

But the question "Is this stuff okay for us to eat?" is as good a beginning as any other, and it has begun to lead to other questions, all of which should become part of a wider discussion by the public, by scientists, and by policymakers. What happens when genetically altered crop plants cross with wild plants? Is it a good use of research money to insert into corn a toxin that the target pest is already developing resistance to? You can't stop evolution — or can you? Is this genetic pollution, and if so, what are the implications of that? Why do we grow corn in monoculture stands that attract diseases and pests? And why do we grow ten billion bushels of corn a year anyway?*

Protesters in butterfly suits make for zippier television coverage, but thoughtful biologists had their own concerns and their own questions: Even though we know quite a bit about the locations of at least some genes and understand some of the mechanics of genetic biochemistry, aren't we still ignorant of the complexities of the processes that biochemistry sets in motion? Hadn't we better learn more about the bigger picture before we start meddling? What gives us the right to meddle? And, perhaps the most important question: Can "big science" (and the science in back of genetic engineering is very big and very expensive indeed) produce good science when it is funded, either directly or indirectly, by agribusiness, pharmaceutical, and biotech corporations that are in a competitive rush for patents and profit? And isn't there something unseemly about patenting the constituents of life?

---

* Thanks to the Corn Refiners Association, which has furnished me with an analysis of the corn crop over the past several years, I have the easy, statistical part of the answer to that question. Currently, about 60 percent of the ten billion bushels of corn goes into livestock feed and another 20 percent is exported. Those parts of the market are flat and can't seem to be expanded too much. But the CRA says that the fastest-growing market for corn is in the remaining 20 percent, which it calls the "industrial sector," and from which it expects future profits. Refinements in the processing of corn as a sweetener were invented only a few decades ago, but in 1985, as cola manufacturers switched to high-fructose corn syrups, corn began to replace sugar cane as the industry's sweetener of choice. A hike down the snack and soft-drink aisles of a supermarket reading labels is a revelation. Corn syrup, in one form or another, sweetens all. The market presents a happy prospect for American agribusiness. As poor people around the world gain a modicum of economic well-being, they can be persuaded, with some marketing effort, to add soft drinks to their diet. All of the six billion people in the world who have not yet done so are potential customers.

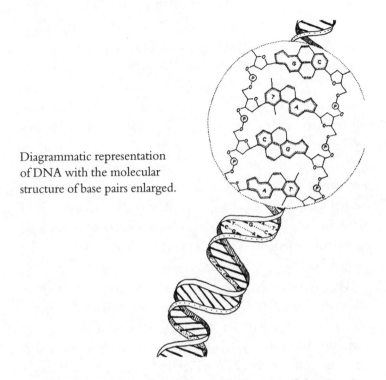

Diagrammatic representation
of DNA with the molecular
structure of base pairs enlarged.

   The furor over the butterflies caught many of the putative
Dr. Frankensteins by surprise, in part because they knew a lot
more about genes and molecular biology than did the rest of
us. They knew, for instance, that genes, which reside on rod-
shaped chromosomes, are found in nearly every cell of plants
and animals. They contain DNA, a big, long molecule made
up of strands of an acid of the sugar-phosphate kind. The
strands are held together by substances called bases, which are
usually designated by their initial letters — A, T, G, and C —
for adenine, thymine, guanine, and cytosine. The bases are
clubby; for chemical reasons they do their work in fixed two-
somes. A and T always go together, as do G and C. These

25

bases bind the acid strands chemically in a ladderlike arrangement that is then twisted into the famous double helix.

Genes are not small bumpy things sitting on chromosomes. They are not different kinds of things in different kinds of life. They are not even "things" at all. They are simply the pattern in which the bases are attached to the ladder's uprights — A-T or T-A or G-C or C-G — and the sequence in which those rungs are repeated and alternated, in functional groups of three, along a stretch of chromosome. When someone "sequences" DNA, it is the arrangement of those ladder rungs that is being teased out. Slight differences in the patterned forms give us the alleles, such as the ones for brown or blue eyes in a human being.

The arrangement of the genes is sometimes thought of as the code, the recipe, for building life's proteins; it is, but genes interact with one another and the organism's environment in little-understood ways. As such they serve as switches that turn on and off certain biochemical processes, that determine what sort of organism a particular bit of life will be, what it will look like, and, to a surprising degree, what it will do and how it will get on in the world.

Each chromosome has many genes — those patterned base pairs of DNA strands. The numbers of chromosomes and of genes vary from plant to plant and animal to animal. A worm whose genome (the total number of genes an organism has in each cell) was recently sequenced had 19,099 genes. As of this writing, the count of the human genome is taking shape, and if the first reports are correct, we may have somewhere between 30,000 and 40,000 genes, a mere doubling of the number for a worm. A better understanding of those "nonfunctional" stretches of DNA may alter that count, but nevertheless it looks to be another reminder that all kinds of life are remarkably similar biochemically.

For our purposes here, however, we can think about the totality of all the genes in all the cells of a plant or animal as a zillion little biochemical factories with switches busily turning on and off all the time. Their sheer numbers, their activity, timing, and interrelationships, the way they modify and change one another — these are the processes of life itself. But no organism lives in an isolation bubble. Other organisms and the environment in which they find themselves affect how those genetic switches are flipped on and off and how the processes work out. Being alive is exceedingly complicated. The really exciting areas of microbiology these days are those devoted to figuring out the interaction of all the parts (mapping a genome takes time but is comparatively mechanical once you learn how to do it, and in that sense the genes are the easy part). An example illustrates some of the complications. Cells store energy to use when the need arises. A plant, for instance, stores energy from the sun, transforming it into a storable molecular form through photosynthesis. Tapping into that energy when the plant needs it is a twelve-step process requiring a different enzyme for each step. Each enzyme is under the control of a separate gene, and each gene must be switched on in the correct order and at the proper time. If that doesn't happen, the plant can't live.

Understanding such chains of events and contingencies is much more than the mere mapping of genes, and it gives a hint of the difficulties facing anyone who attempts to solve an agricultural problem by putting a gene from one organism into the genotype of another. An acquaintance of mine — an eminent medical researcher and Nobel laureate — told me privately of a further difficulty . . . and an economic reality. His interests in genetic engineering lie not in what we do with other animals or plants but in what we do with ourselves — in short, genetic "therapies" for human ills. He worries that this notion is being oversold to the public. It is true that some human diseases are caused by a single defective gene and might respond to genetic treatment, but those diseases are so rare that medical technology companies would not make a profit from treating them. Instead, what those companies want to treat are the more widespread and potentially profitable human ills. "But those are not caused by a single defective gene," my friend said. "They may have a multigenic basis that can give a person a predisposition for the condition. However, among people with that sort of genetic profile, some develop the condition and some never do. Perhaps it has something to do with the way they live. There are better ways to treat those conditions than with gene therapy, and what we need to know more about is what makes some people vulnerable and others not."

Genes can be seen as a code written out in four letters or a tune played on a four-note theme in triplets.* Each person, each stalk of corn, each dog has a tune that is different from that of any other person, stalk of corn, or dog. But within a given group of plants or animals there is a characteristic tune — that is, the stretches of patterned ladder rungs are much

---

* I am in intellectual debt to Richard Powers, who, in his brilliant novel *The Gold Bug Variations* tells the story of a biologist who explores the structure of DNA using Bach's music as a model.

the same within a species. And it is surprising, at first, to see how much alike the pattern is from one species to another. It is currently thought that the similarity of those four-note variations of life's theme of A-T, T-A, G-C, and C-G shows how closely related one group or species is to another. The patterns worked in our DNA, for instance, look very like those of chimpanzees and enough like those of dogs or cats that genetic experiments conducted on them pertain to us. And our molecular construction turns out to be more like that of yeasts than anyone would have imagined fifty years ago.

This similarity is not really surprising, however, because the evolutionary process is a ferociously conservative one. Why reinvent the wheel? If some pattern in the DNA works, why change it? Creation is a stingy affair. New species come about through modification of existing sequences of the pairs of As, Ts, Gs, and Cs and through the addition of new patterns as well. And the old sequences can make do in new ways. For instance, flies and moths diverged from one another about two hundred million years ago, yet when researchers identified several genes responsible for creating wing spots in one butterfly species (butterflies are lately specialized moths), *Junonia coenia* or buckeye, they found that the genes had the same sequence as those that are central to wing formation in the fruit fly. The thorough study of the genetics of fruit flies has illustrated this basic conservatism of genetic structure. A genetic alteration that makes a fruit fly learn more quickly can also enhance learning ability in snails or even mice, animals that are, genetically speaking, close relatives of humans. It has been six hundred million years since our ancestral stock and that of a fruit fly parted company, but of the 289 mutant genes known at this time to cause disease in humans, 177 have direct counterparts in the fruit flies.

All the patterns that make up life on the planet are variations on the same four-note theme, and all can be seen, in their simplest terms, as embodying biochemical processes that are universal, despite the complexity of the flourishes and trills. Buried in the media stories about *Bt*-corn was the biological reality of this universality, which is what makes transgenic engineering possible. Biology has told this story many times and in many other ways. Considering the public reaction, however, perhaps it needs to be repeated: at its biochemical base, life, in whatever form it takes, is pretty much the same. Corn and *Bacillus thuringiensis* are not all that different, nor are we. If that is news, it is one of the cheerfulest pieces of news I know: all of us — corn, humans, dogs, yeasts, fish, tobacco, and bacteria — are fellows one to another. We are together in this matter of life.

When biologists talk about genetic distance, they mean the extent of variation from one kind of life to another. The new tools of molecular biology, which are better able to examine genetic distance, have upset old ideas about what a species is. Even though many of us may think of "species" as a fixed and forever category of reality, biologists know better. They regularly revise and reorder and rename species and even debate the definition of the term. ("Genus," the first word in a scientific name, lumps together related species. "Species," the second word, stands for uniqueness. And that uniqueness is what biologists are increasingly having trouble defining.) One example can suggest the kinds of puzzles biologists are considering and why they make the shift of a gene from one kind of life to another not quite so startling as it might seem at first.

Linda Maxson, a dean at the University of Tennessee, was reflecting on her life as a biologist when she wrote, "I studied two salamanders. They were taken from under the same log. They looked identical. And the genetic distance between them was larger than that between a human and a chimpanzee [chimpanzees are one of the apes with whom we share 99 percent of our genes]. Such experiences lead us to reexamine what a species is." This example is not an uncommon one for biologists, and it makes them ask one another whether two identical-looking cohabiting salamanders represent one or two species.

Are human beings and apes different enough to be considered more than a single species? Of course they are, we all reply. Apes and humans look different, do different things, and have different capabilities. Even a child can see the difference, knows that apes are the ones behind the bars at the zoo and people are the ones on the outside. Children know that because we tell them so, just as we have been told. Seeing is shaped by knowing. How would we see an ape if we had not been told what it was?

It so happens that we have a record of such an experience concerning the animal we call gorilla, another one of the apes with whom we share almost complete genetic likeness. The word "gorilla" is derived from an unknown African language as heard by the Phoenicians, the first Mediterranean people to see the ape and the first historical people to circumnavigate Africa. The explorer Hanno, who sailed along its western coast, had with him some captured local people to explain what he was seeing. A document in Greek, purported to be a translation from a Phoenician inscription written by Hanno, has been handed down. In part it says:

> In the recess of this bay there was an island full of savage men. There were women, too, in even greater number. They had hairy bodies and the interpreters called them Gorillae. When we pursued them we were unable to take any of the men; for they all escaped, by climbing steep places and defending themselves with stones; but we took three of the women, who bit and scratched . . . and would not follow us. So we killed them and flayed them, and brought their skins to Carthage.

A number of writers from later times, but before the Romans burned Carthage, reported having seen the skins of those women, who were said to live in the south. Some even said that they were the Amazons, those fierce women warriors.

Does the category "species" have any meaning? We still need a word of that sort and biologists still use it, but to them the term "species" seems much looser and a little more slippery than it does to the public.

Botanists, especially, have long been uncomfortable with the old species definitions, which, they believe, have been dominated by zoological thinking, based on the interbreeding of animals to produce offspring like the parents. Even the more recent and hedgier definition in my biology reference book, which calls a species "the largest unit of population within which effective gene flow occurs or could occur," can be a problem for botanists, because the idea behind the word "unit" doesn't always match botanical reality. For example, cottonwoods and balsam poplars separated from a common ancestral stock twelve million years ago, and they are recognized as separate species. Yet they mate easily and produce fertile crosses. Or take dandelions, which gave up pairing long ago and reproduce asexually. They still produce pollen, but it is sterile. Genetic flow is strictly from a single parent dandelion to its offspring, which grow up from those fluffy seeds that waft with the breeze. Does that mean that each maternal line of dandelions — the ones in my yard and the ones in yours — is a different species? Is that a useful label? Is it absurd? Those are just a few of the problems botanists deal with in identifying species.

In addition, those involved in agronomy and horticulture, in particular, are perfectly comfortable with species that *are* penetrable, loose, and unbounded. Horticulturists work with plants that are a veritable genetic hodgepodge of crosses and hybrids, and they use the word "species" for a group of plants that are horticulturally a unit but that have parents from different species. To indicate this they use an X between the genus name and the species name. Apples, which we'll meet up with in Chapter 4, go by the scientific name of *Malus* X *domestica*. That means that the apple from the supermarket in your brown paper lunch bag is a cross. It has the genetic makeup, actually, not just of two disparate parents but of a whole clutch of disparate ancestors. And if you were to plant a seed from your lunch apple and let it grow into a tree, the fruit of that tree would neither look nor taste like the apple the seed came from.

All in all, the scientists in white lab coats whom the public was calling Dr. Frankensteins were startled about the reaction to the *Bt*-corn story. Many of them had their own doubts about genetic engineering, had their own questions to ask, but they knew that inserting a *Bt* gene into corn was not nearly as big a deal as creating corn in the first place. *That* was a very big deal.

There is more to this story, however, than corn, which is what caught public attention, courtesy of the monarch butterflies. And there is more to it than food. We have created many other new species in addition to corn. And we have fiddled with the genetic makeup, the biological identity, of all the kinds of lives with which we have associated. Artifice *is* our nature. But this has not been a one-sided project. The plants and animals that we have made or rearranged have given human history a push here and a nudge there and sent it off in new directions. The story is best told by taking a look at what a few of those plants and animals, fellow travelers of ours, have meant to us and what we have meant to them.

Of Multicaulismania,
Silkworms,
and the World's First
Superhighway

But Silke (whereon my loving Muse now stands)
Was it the offspring of our shallow braine?
— Thomas Moffet, "The Silkwormes and Their Flies," 1599

BACK IN MY DAYS as a commercial beekeeper in the Ozarks, I had a beeyard on a farmer's land near an old house, long since abandoned by its owner. From its style it appeared to have been built in the late 1800s or early 1900s. That was a time when a considerable number of families came from the East with their grubstakes and hopes, to homestead after life's problems had discouraged them elsewhere — or simply to find adventure. I set up ten hives behind the farmhouse, where they were protected from the winter wind by a row of aged mulberry trees. The bees gathered pollen from the tiny white blossoms in the springtime

OPPOSITE: Empress Hsi-Ling Chi with a cup of tea, by means of which she invented silk culture. Inspired by an eighteenth-century illustration by Hokusai Katsushika.

but never took any nectar from them as far as I could tell. Later in the summer the trees' branches drooped heavily with fruit that looked like giant blackberries. They weren't as good to eat as blackberries, but I nibbled on them occasionally anyway.

The trees were white mulberry, *Morus alba,* rather than the native red, *Morus rubra.* White mulberry comes from Asia, where *Bombyx mori,* a species known as the silkmoth, was created. The white mulberry is the silkworms' food plant. It is impossible to know what dreams and hopes the family that built the farmhouse brought with them, but the presence of those white mulberries suggests that they may have fallen under the spell of multicaulismania, as it came to be called — the dream of growing rich by growing silkworms on the variety of white mulberry known as *Morus multicaulis.* Raising silkworms was the rural get-rich-quick scheme of its times, like llama ranching, Vietnamese potbellied pig husbandry, designer vegetable farming, earthworm farms, pick-your-own strawberry patches, or mail-order herbs in recent years.

As an agricultural animal, silkworms have been a part of government economic policies right from the beginning, five thousand years ago. And that was how they, along with their food plant, the white mulberry, ended up in places like Missouri when it was settled.

England, yearning to have a slice of France's silk industry, had long tried to establish silkworms at home, but the climate was not suitable. As America began to be settled, colonial chartering bodies viewed the new continent as a splendid place to produce all sorts of raw materials for England's factories, silk fiber among them. The Virginia colonists in 1619 were required to raise silkworms. After all, the red mulberry tree grew wild in America and the worms ought to be able to learn to eat its leaves as well as those of the white mulberry, which was more demanding to grow. Silkworm eggs were handed out for free, and a fine was levied if the colonists did *not* raise caterpillars from them. A nameless publicist composed an exhortative jingle:

> Where Wormes and Food doe naturally abound
> A gallant Silken Trade must there be found.
> Virginia excels the World in both —
> Envie nor malice can gaine say this troth.

That mercantile experiment failed, just as a Spanish government program had a century earlier in New Spain (Mexico). In 1522 Hernando Cortés appointed government agents to bring mulberry trees and silkworm eggs to the new colony, but by century's end all traces of both were gone.

Silkworms are tricky animals to raise, and the failure of the Virginia colonists to create a silk industry probably had many reasons, but for starters, silkworms, *B. mori,* do not feed on *Morus rubra,* the red mulberry. (Later it was discovered, in the search for a substitute food, that the caterpillars would eat lettuce and the leaves of Osage orange, a tree native to the south-central United States, but they do not thrive on those leaves, and no silkworm industry was established using them for fodder.)

But England did not give up on making its new colonies produce silk. During the late 1600s and early 1700s it began sending white mulberry seed and seedlings to be established in nurseries on Long Island. An early president of Yale, the Reverend Doctor Sayles, caught the silk fever and tried to introduce silkworm cultivation into Connecticut before the Revolution. In Philadelphia, Benjamin Franklin was trying to start a filature, as the place where silk thread is reeled from cocoons is called, when the war broke out. And after the United States became independent, Congress took a great interest in furthering the silkworm industry, passing the first act for its encouragement in 1828. The policy was a form of social engineering as well as of economics, for it was hoped that silkworm growing would become a cottage industry, the sort of thing women and children ought to be good at, which would make the citizens of the new country prosperous.

But despite encouragement from high places, despite the bounties that were offered, no American silk industry emerged. Enticed into the enterprise by promises of high income, householders soon discovered that the problems were many and the profits, when there were any, were too slight to warrant the time required to nurse the fussy silkworms. After caring for them for an entire season, they might, on occasion, harvest enough fiber for a pair of stockings or a single glove. As always, however, some people *were* making money: those selling supplies and equipment, silkworm eggs, and mulberry trees did quite well for a time. Even in its earliest days in America, silkworm growing was, in effect, a scam, for the only ones to make a profit were speculators who supplied the wherewithal to those who did the work.

Nursery owners began to look for a better white mulberry that would make the silkworms more productive. In the 1830s they thought they had found one: *M. multicaulis* (now considered a variety of *M. alba*) was a mulberry from the South Seas that grew extra-large leaves and appeared to thrive in the United States. People's get-rich-quick hopes were raised again. Speculators sold land that was to be cleared and planted with thousands of *M. multicaulis.* The trees, which first sold at four dollars a hundred, jumped in price in two years to thirty dollars a hundred. In one week in Pennsylvania, three hundred thousand dollars changed hands for a single batch of shoots, which were sold over and over again at advancing prices. A few people grew rich before it was discovered that *multicaulis* was sensitive to frost. By the end of the decade, that particular boom had collapsed and the *multicaulis* plantations were uprooted.

Agronomists shook their heads over multicaulismania but did not lose faith in the enterprise. In 1858, ten years after gold was discovered and gold diggers' hearts were broken in California, the state government began offering bounties to encourage new settlers and disappointed miners to establish white mulberry orchards. Thousands of trees were planted, and the wily tried to collect the bounties even without planting them. One grower demanded fifty thousand dollars from the fund for what he said were one million mulberry trees planted in two hundred plantations. But government inspectors, checking out the increasing reports of fraud, discovered that he had planted only a couple of acres with seedlings and that they were planted too close together to ever grow. His claim was dismissed.

In Utah, Brigham Young ordered trees and instructed Mormon women to spin silk, which they did diligently for a time. But once again the profits came only from the sale of trees and silkworm eggs.

Still, the U.S. Department of Agriculture, always and forever a cheerful agency in the face of dismal facts, did not admit to any doubts about this departmental pet project. The report of the commissioner of agriculture, Horace Capron, in 1868 begins with the statement: "That the culture of silk can be profitably carried on in the United States is clearly established." Capron goes on to analyze the $215 million world trade in silk (the experienced Chinese were responsible for over $81 million of it), the tantalizing interest in American silkworms shown at the Paris Exposition, the equally tantalizing fact that silk was selling for its weight in silver, and that France obtained seven times the revenue from silk that Mexico did from its mines. Then he slips into a delirium of federal prose:

> An acre should support from 700 to 1,000 mulberry trees, and, when four years old, they should produce 5,000 pounds of leaves to the acre; that is 4,000 pounds suitable for feeding, and, during feeding time, without injury to the tree. Those leaves should feed at least 140,000 worms, which will produce 70,000 female moths, and those will lay 300 eggs each, or 21,000,000 in all. After deducting 5,000,000 for possible loss, we have 16,000,000 eggs, or 400 ounces for sale, or $1,600 per acre . . .
>
> A Californian, travelling in Europe, had taken a couple of . . . cocoons with him; and in Lyons he showed them to a manufacturer, and wanted to know the value of such cocoons by the quantity. Before he had finished his sentence the Frenchman

seized them, called his friends to admire them, and said that the country which could produce these in quantity had better wealth in its valleys than in its mines. Europe had nothing to equal them.

The only problem (the commissioner should have read *Jack and the Beanstalk*) was that it didn't work out that way. Like hopeful hordes before them, Californians discovered that silkworms are not easy to raise and die in droves when their exacting requirements are not met. By the late 1870s, the western silk craze had met the same fate as the earlier *multicaulis* mania. One reason that the French manufacturer had shown such an interest in the Californian's silkworm cocoons was that France, which had its own sericulture industry, centered around Lyons, was having a good deal of trouble with its silkworms. Silkworm growing had proved more successful in France than it had in the United States, but simply for that reason there was a large concentration of silkworms in the region around Lyons. A large concentration of any particular organism, be it corn or silkworms, becomes attractive fodder for the organisms that prey upon it. The silkworms of Lyons during the 1860s suffered from a disease called pebrine, which caused the eggs to hatch out as stunted, spotty caterpillars that were poor feeders and languid spinners of loose cocoons. Their filaments produced an inferior thread. The French, and others, were looking for a better silkmoth than *B. mori,* one that was not susceptible to pebrine. Léopold Trouvelot, a French astronomer who had interests in natural history and had moved to Medford, Massachusetts, was one of those who was trying to develop a better silkworm. He experimented with several American moths that produce silk, such as the big, pale green luna, the prometheus, and the cecropia, but he discovered that none of them was suitable for silk production and that they could not be domesticated. After giving up on hardy native American moths, Trouvelot turned his attention to some European species and imported to this country from France a moth identified in those days as *Bombyx dispar,* whose adult form looked something like the silkmoth, *B. mori,* and seemed to belong to the same genus. *B. dispar* was a tough, feisty moth that had moved steadily westward from Eurasia across the Continent, arriving in France hundreds of years before. As it came into new territory it caused a certain amount of trouble because in its caterpillar stage it was a voracious and catholic feeder. Not at all picky about the kinds of leaves it fed upon, it often stripped trees bare. But after some few generations the predators that fed upon *B. dispar* moved in, too, and kept it in check. In England it was such a rarity that moth collectors paid enormous sums for specimens of it.

Trouvelot, accepting the classificatory wisdom of the time, which said that *B. dispar* belonged to the same genus as the silkworm, seems to have hoped that he could crossbreed the two moths, giving the offspring some of *B. dispar*'s toughness and exuberance while retaining the silk-producing qualities of *B. mori.* In 1869, in the course of his experiments, some of the *B. dispar* cocoons were knocked out of a window in his lodgings, and the moths that emerged from them spread zestily into a new environment. That was unfortunate; today's taxonomists know that those moths do not belong to the same genus as the silkworm, or even to the same family.

Today they are known as *Lymantria dispar*. *Lymantria*, meaning destroyer, is a more appropriate name, for *L. dispar*'s common name is the gypsy moth. It was to fight the national infestation of gypsy moths that the "natural" pesticide *Bt* was developed after the public outcry against chemical insecticides.

By the end of the 1800s Trouvelot had given up his crossbreeding experiments for astronomy, and the California silkworm boomlet had fizzled, but the USDA never gave up hope of establishing a silk industry in this country. It built a state-of-the-art filature in the Agriculture Department building in Washington and bought cocoons from producers. In 1887 the USDA reported that the entire output of the filature brought only $865.81. Government figures also show that around this time more than one thousand people were growing silkworms. Nine of them, the government report proudly states, "made more than $40 for their work." With federal encouragement "a number of women selected from among the more skilled silk growers [went] from place to place [to] teach the reeling of silk to children. They received no salary, and their only compensation was free board from the farmers."

In 1901 Congress appropriated ten thousand dollars to encourage silk culture in the South "to . . . ameliorate the condition of the extremely poor people . . . particularly of the colored race."

Leland O. Howard was the author of the above quotations, which are taken from his report in the 1903 yearbook of the Department of Agriculture. It is obvious that Howard had acquired the silk fever. With a thoroughly American attitude, he viewed the chief hindrance to a thriving sericulture industry as the lack of good machinery. If better machines were developed, he argued, the big factories that would be built to house them would bring about economies of scale. The cost of finished silk would be cut, and it could be sold competitively on the world market. In a passage that sounds as overwrought as the 1868 report, Howard wrote:

There are many portions of the United States well adapted to silk raising, many places which might become silk centers, where labor can be employed practically at rates comparable to those of southern Europe. The establishment of a silk mill in such a location, with its own filature attached, with the surrounding people employed as operatives in both the filature and mills, and with the otherwise unoccupied members of their households engaged in silk raising in the spring, is feasible, and can be made to pay. A beginning of this kind may possibly soon be made by foreign capital. The proprietor of a large estate in Italy is at present giving the matter serious consideration. A foreign proprietor of a silk establishment in one of our northern cities states that he can count upon the employment of 5,000 of his compatriots more or less skilled in the silk industry at an average daily wage from 20 to 25 cents.

To further the dreams of Howard and other federal policymakers, the USDA printed encouraging booklets about silkworms for public distribution. The government was also in the business of buying and distributing white mulberry trees — but of a hardier strain than *multicaulis*. It also distributed, for free, silkworm eggs to anyone who wanted to grow them. There remained, however, a pesky problem: there was no market for the American cocoons, despite Howard's hopes.

46

So in the early years of the twentieth century, long before New Deal economic policies, the USDA was buying cocoons at European prices from American growers and reeling them at its own filature. There is no record of any commercial success in the enterprise, which produced at best only a few hundred pounds of silk a year.

Even in 1903, when Howard wrote his optimistic report, Europeans were developing "artificial" silk, and Americans soon found the synthetic fibers derived from cellulose easier and surer to produce than silk had been. In 1924 the U.S. Department of Commerce officially dropped the name "artificial silk" and declared the fiber to be "rayon." I wonder whatever happened to the USDA filature.

I also wonder what hopes had been invested in those white mulberry trees at my Ozark beeyard. That particular area was settled around the turn of the century. Did the people who homesteaded there arrive with the government pamphlets on silkworms in hand? Did they bring the young trees, packed in sawdust, in their wagon? Did they plan to have a few chickens, a cow, and a pig or two, put in a garden, and do something . . . perhaps silkworm growing . . . for cash for coffee, sugar, and taxes? That was the way later settlers arrived, the city folks fleeing the Depression in the 1930s and the back-to-the-landers in the 1970s. After rayon and other artificial fibers relegated silk to the status of a minor luxury product, silkworms no longer figured in the plans of many homesteaders. By the 1970s, when I arrived in the Ozarks, everyone was talking earthworm farms. But the dreams, I suspect, were similar. I hope that whoever homesteaded that spot where the mulberry trees grew had a good run of time in the Ozarks. It was a fine beeyard and I always liked the place.

There is a minor coda to this American part of the silk-worm story, a story that otherwise takes place in the Far East and elsewhere. In the 1980s a West Coast art gallery was dis-playing the work of Kazuo Kadonaga, a Japanese conceptual artist, who briefly made silkworm art popular in the United States. In his first pieces Kadonaga simply turned silkworms loose on wooden grids similar to those used by commercial growers, allowed them to spin their cocoons as they chose, and called the final product art. But the silkworms, indiffer-ent to the artist's sense of design, bunched up and tended to pupate doubly or even triply within certain boxes in the grid, leaving big sections bare. Kadonaga then enlisted the help of an assistant, and together, for the hours during which the ripe caterpillars tried to settle down and spin their cocoons, the two men rotated the grid constantly, tricking the silk-worms into acting as though down were up and so was sideways. Wearied, and perhaps confused and dizzy, the silk-worms eventually settled down in a random, more aestheti-cally pleasing manner throughout the grid, some singly, some doubled into the same grid cubbyhole. When they were done spinning, Kadonaga heated the grids to kill the cater-pillars within the cocoons. The gallery sold many of the pieces. A friend gave me one, and it sits on a shelf behind my desk as I write. It is framed in painted wood and sealed tightly in glass to keep out the bugs that would otherwise infest it and the mice that would chew the cocoons. The sharp angu-lar forms of the grid contrast with the curves of the inch-and-a-half-long silken cocoons and the graceful swooping

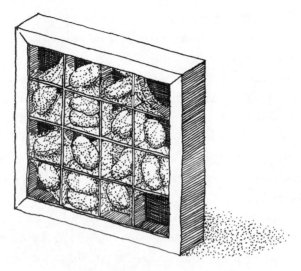

My piece of silkworm art
by Kazuo Kadonaga and the silkworms.

threads of silk that the caterpillars trailed across the grid in their search for a proper spinning home.

Kadonaga had an international reputation by the time his work was shown in the United States, and a well-known art critic writing in a prestigious art journal reported on the silkworm art show, calling it a "complex interplay between man-made and the natural." I assume he meant that what the silkworms had wrought was natural and that the concept behind the pieces, as well as the wooden grids, the painted frames, and the glass, were man-made. But the concept, the milled wood, and the glass are just as much our natural products as the silkworms' fluffy cocoons are their natural products. Artifice *is* our nature. And the irony is that the worms, with their silken thread, are also man-made. Silkworms, as they now exist, are a human artifice, a product of our nature, not theirs.

The double-cocooning within the grid suggests that the caterpillars might have been F_1 hybrids of Japanese and Siamese strains of silkmoths, which were developed in the early 1900s. Double-cocooning is considered one of the faults of that strain. It is a fault because the two cocoons become intertwined, making their silk hard to reel. All of these "industrial strains," as the hybrids are called, have been so genetically modified over the five thousand years of their cultivation that *Bombyx mori* is now biologically distinct from its wild progenitors. And that genetic modification is of our doing.

Many moths (and other insects and spiders, too) spin threads, or filaments, as they are more properly called, and people have used many of these at one time or another to make silken thread. Even in the New World in the time of Montezuma, silk was a trade good among the Aztecs. It was made from the cocoon of the hammock-net moth.

In Mediterranean countries, long before the first millennium of the Western era, Aristotle and Pliny both gave garbled accounts of a fabric made by women on the Greek island of Cos. They claimed that the "butterflies" there made "wool" by scraping the down from leaves, which they wrapped themselves in for protection from the cold. The women of Cos gathered the wool and spun thread from it. Irene Good, a silkworm archaeologist at Harvard University, biochemically analyzed those ancient silk fibers from the Mediterranean world and discovered that they were spun by the Cos silkmoth, *Pachypasa otus,* which is native to the area. A big moth, with zigzag markings on its forewings, *Pachypasa* belongs to the same family as the tent caterpillars that spin fuzzy, webby nests in wild cherry trees in this country, but the Cos silkworm feeds on pines and other evergreens. It produces only modest amounts of silk, and reeling thread from it and weaving cloth must have been a lot of work.

Cocoons of several moths of the saturnid family, commonly called giant silkmoths, have been used, and still are in India, to make a coarse silk. Irene Good suspects that Indians had learned, as early as the Chinese, the real secret of silk making, which is that before the cocoon filaments can be reeled, they have to be degummed to remove the substance that glues them together. In the true silkmoth of the *Bombyx* genus, the caterpillar must be killed as soon as it has finished making the cocoon around itself, for if it completes its transformation and comes out of its silken nest, it will break the filaments into short bits that cannot be reeled. According to Irene Good, widespread religious taboos against the taking of life, even that of a moth, prohibited Indians from using silk from *Bombyx* moths, so they developed very early a silk industry employing the saturnid moths of the *Antheraea* genus (our Polyphemus moth, which Trouvelot experimented with, is a member of that genus). Those caterpillars conveniently leave a small opening at the end of their cocoons when they spin them, and the adult moths can escape through this hole without breaking the filaments. Those moths resist domestication, however, and their cocoons had to be hunted down in the wild.

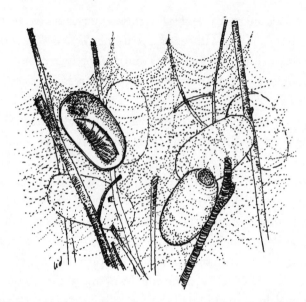

Silkworm cocoons on traditional twig supports. One is cut open to show the chrysalis within. An emerging mature moth has made a hole in the end of another cocoon, breaking the silk filament.

So it was left to the Chinese to develop one of the greatest trade monopolies the world has ever known, one that would bring East and West together. The Chinese knew no taboos against killing moths. They developed silk in quantity as a trade good and in the process they developed a new species of silkmoth.

According to legend, it all began when Hsi-Ling Chi, the principal wife of Emperor Hwang-ti, was walking in her garden in the province of Shantung one fine afternoon in 2640 B.C. She noticed that caterpillars were no longer crawling about on her favorite mulberry tree, and in their place were fuzzy cocoons. Hsi-Ling Chi picked one off to look at it and quite by accident dropped it into the cup of hot tea she was carrying. When she took it out, the delicate fiber of the cocoon unraveled into a long thread. Having thus discovered the secret of loosening the gummy sericin (it wasn't the tea specifically — any hot liquid would have served), the empress was admitted to the company of the gods. Excited with her discovery, she showed the thread to her husband, the emperor, and he ordered a robe to be woven from it. Having no qualms about boiling caterpillars, Hsi-Ling Chi devoted the rest of her life to the culture of mulberry trees, the nurture of silkworms, and, it is said, to the invention of the loom.

Later empresses had the ceremonial duty each spring of beginning the silkworm rearing that was carried out by peasant women. The government took a continued interest in silk production by setting out sacrifices, including human prisoners, at the altar of the Silkworm Goddess under the direction of the Mistress of Silkworms.

The wonderful specificity of Hsi-Ling Chi's story notwithstanding, sericulture was well established by at least 2640 B.C. Silk fragments, dating to between 2850 B.C. and 2640 B.C., have been found in good enough condition to analyze microscopically, revealing that the silk in them was spun by *Bombyx mori*. The filaments of each silk-spinning moth, and even of each race within the species, have a characteristic structure that can be seen under a microscope.

★

I spent a snowy winter day in Rhode Island not long ago with Marian R. Goldsmith, one of the few people working on *B. mori* silkworms in the United States. A professor of biology at the University of Rhode Island in Kingston, Marian is a vivacious, diminutive woman with a reputation in biological circles out of all proportion to her size. She is widely known to entomologists as the editor and a contributor to a text that sets forth an important new way to study the genetics not only of moths but of other insects as well. Although she spends much of her time in this country, she is the American member of a team of academic and government geneticists in Japan doing pioneering work on genetics using the silkworm as a model.

We sat in her cozily cluttered office with its stacks of books and overflowing files. She pointed out a glossy Christmas card on her wall from a French silkworm research institute. It featured a photo of a glowing green silkworm on a cheerful red background, the glow courtesy of a bit of transgenic engineering involving jellyfish genes, a technique that has recently produced a glowing rabbit and chimpanzee. This makes for a startling-looking animal, but the real reason for the transgenic feat is that by adding a fluorescent marker gene to the animal, researchers can tell if a given gene transfer has worked.

The dimensions and character of silk are under genetic control. The genes governing the shape of the worms' silk spinnerets can be changed, and this in turn changes the silk that comes from them. Marian told me of one special race of *B. mori* developed in Japan. Certain Japanese musical instruments use strings that must be wrapped in silk of a precise texture and dimension, and a particular race of caterpillars has been created to spin this silk.

Scientists believe that the species *mori* was created, long, long ago, by genetic modification through breeding of moths of the *mandarina* species. *B. mori* has the same number of chromosomes — twenty-eight — as several races of *B. mandarina,* and in China, where those races of *mandarina* live, the greatest range of genetic diversity in *B. mori* is found. This diversity would indicate that China was where *B. mori* originated. Because *mandarina* is a wild moth, its pace of evolution probably was slow, resulting from the sort of genetic change that made it better suited for its wild environment and assured the survival of its young. *B. mori* would have changed more rapidly, through inbreeding, controlled crossbreeding, and the preservation of mutant forms.

B. mandarina is a small moth that flits nervously in flight. Its mottled brown and cream coloration helps it hide in the dappled shade of a mulberry tree, and it sometimes aligns itself with the ribs of leaves for further camouflage. Its caterpillar roves widely in search of mulberry leaves. It wanders while it gets ready to spin its cocoon and takes several days to settle into its chosen location and complete the spinning. Its cocoon is a loose, untidy affair that would be difficult to reel.

In contrast, *B. mori* has become a huge, heavy-bodied moth that is placid in disposition. Its size allows it to spin enormous amounts of silken filament, but it is so heavy that it can no longer fly, though it still has wings.

Just about the first rule of domestication is that the animal

Adult silkmoths on a mulberry leaf.
Left, *Bombyx mori;* right, *B. mandarina.*

you are to keep must not run away. *B. Mori's* heaviness may have begun as a chance mutation or an expression of some hidden allele. These larger moths might have been kept for breeding, but perhaps not primarily because they produced more silky filament. The early archaeological evidence of the species is as cut-open cocoons, and while experts disagree on what that means, some suggest that people may have been eating the pupae (people eat larvae in some parts of the world today). The biggest and juiciest ones would have been the most desirable, of course.

Present-day *B. mori* moths are much creamier and lighter in color than *B. mandarina,* and their caterpillars have lost

B. mandarina's twitchy, wild way of feeding. They do not hide nor do they roam while feeding. They simply wait patiently to be given food by a human hand, and they placidly die if it is not forthcoming. All of those qualities would make them visible and vulnerable in the wild, but of course they do not live in the wild.

In certain regards silkmoths are the zoological equivalent of corn. Both species are of human devising, and both are incapable of living on their own in the wild. Silkmoths, too, would be missing if humans returned to earth after five thousand years of captivity by little green men in spaceships. They are arguably the most completely domesticated and genetically modified animal we have. We created them, and they cannot live without our care.

Raising silkworms and reeling their silk was women's work in ancient China, a cottage industry supported by the government. Under Emperor Yu (2200 B.C.) huge tracts of land were drained for the planting of mulberry trees, and silkmoth eggs were distributed free of charge to cottagers. The earliest silk thread, the product of their labor, became the exclusive property of royalty, who had it woven into lustrous, soft clothing in patterns and colors that only they could wear. As more productive silkworms were bred and the peasantry became more adept at raising and reeling, the privilege of wearing silk was extended to the nobility, with specified colors and patterns for each rank. Silk was such a beautiful, much-wanted fabric that sericulture became required serf labor.

Those serfs must have been very skilled indeed, for they

eventually produced so much silk that even the less-than-noble classes could wear it. Rumor of the luxurious, lustrous cloth spread through Asia, and probably some silk was smuggled out of China. The abundant supply and a period of hard domestic economic times finally transformed silk into a trade good. A state-sanctioned silk trade was established in 200 B.C., but the government maintained a lucrative monopoly by tightly holding the secrets of its production. Anyone who revealed them was, if caught, promptly beheaded.

Nevertheless, such a beautiful and lucrative luxury good could not be kept secret. Silk was known to the Parthians,

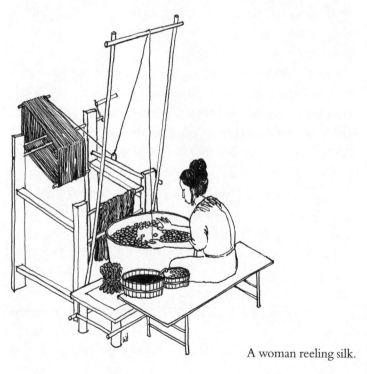

A woman reeling silk.

ancient inhabitants of the land southeast of the Caspian Sea, part of which today is Iran. They are thought to be descendants of the Scythian nomads who roamed Central Asia. In the second century B.C. the Parthian king Mithridates encouraged trade with the East, and the Parthians once used silk as a secret weapon to defeat the Romans. In the midst of battle, at noon on a sunny day, the Parthians unfurled silken banners that shimmered and shone. The Roman soldiers had never seen such a sight (or so the story goes), and they retreated in confusion.

The Mediterranean world had known of the countries to the east at least since Alexander invaded India in the fourth century B.C. But formidable mountains, deserts, and warrior tribes separated the high culture of the East from the high culture of the Greeks and Romans. And for peoples of high culture in between, such as the Parthians, there was good reason to keep East and West separate: it allowed them to monopolize the trading profits. But the Mediterranean people learned that an eastern people they called Seres — and we call Chinese — made a much finer, more shimmering fabric than any made by the women of Cos.

The Seres, in turn, had heard rumors of people living to the west in a land they called Li-Kan, the wide swath of territory between Parthia and Rome. They were eager to have what they called the "heavenly horses" from the "western Regions," by which they meant Bactria in modern Afghanistan, and they may have traveled westward to trade for them long before the new millennium began. Once suitably garrisoned in Bactria against bandits and warlords, the Chinese

SILK ROADS

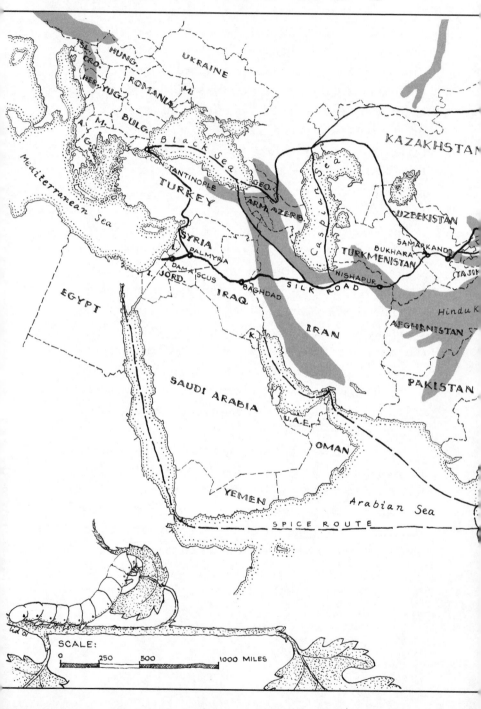

SCALE:
0   250   500   1000 MILES

traders could sell silk and buy furs and woolen carpets from Central Asia, pearls and spices from India, and Mediterranean glassware.*

The Western traders carried their goods — tin and glass, for instance — from port cities such as Constantinople and Damascus as far east as Persia, where they exchanged them for silk. As it traveled from China to Europe, the silk became very expensive, but the traders found a good market for it in the Mediterranean world. The rich and powerful were eager to drape themselves in silk, for its very price spoke of luxury and privilege.

Despite the growing exchange of goods, the Romans and the Chinese did not have any real contact with each other. Those fierce Parthians (the ones who had flummoxed the Romans with their waving silk banners) were the go-betweens in all the trading. Eager to keep things that way, they told tall tales to both East and West about how difficult and dangerous the route was and how untrustworthy were the customers at either end of it. In their glory days, the Parthians controlled not only much of Persia but also Assyria and Macedonia. One of the major way stations along the Silk Road — the five-thousand-mile stretch of highway between the Mediterranean and China — was Nishapur, not far from the modern Iranian city of Mashad. And indeed, just to read through the lists of trade goods that passed through a

---

* The West's love of silk and ignorance of its source make it seem as though the Chinese had the better end of the trade. But they, in turn, were ignorant of glassmaking and took Western artisans' glass beads as precious gems, believing, at least for a time, that glass was crystal. And who were the Parthian traders to disabuse them?

city like Nishapur is to wallow in sensuous opulence: lapis lazuli, topaz, turquoise, ivory, and onyx. Frankincense and spices. Silver, coral, wine, dates, and peppers. Camellias and cats. Slaves and safflower. Peaches, apples, chives, coriander. And on and on and on. And silk, always silk: silk thread, silk woven into fabrics, silk draperies and carpets. Silk. Silk. Silk. Trading was so lucrative that some of the goods stayed with the merchant traders along the route, who had become rich and powerful themselves and had a taste for the finer things. And, in due course, when they were able to winkle the secrets of silk making out of the Chinese, a famous silk industry grew up in Persia itself.

The Silk Road was no nicely marked interstate system with green reflective signs and convenient rest stops. It was, actually, a series of routes, roughly parallel, and the decision to travel on any particular one depended as much on who was on the political outs with whom as it did on the terrain or weather. But the Parthians' horror stories notwithstanding, the Silk Road did offer some conveniences that sound quite modern. Chinese caravan travel through the Tarim Basin, a barren wilderness of saltwater swamps in western China north of Tibet, was made easier by the laying of pipes carrying fresh water to travelers. And, like today's truck drivers giving one another tips via CB on road conditions and particularly good pie at a given truck stop, Parthian merchants exchanged travel information about routes and way stations. Some of that information was written down in early versions of today's travelers' guidebooks. One that survives to this day, entitled *Parthian Stations,* gives the locations of caravansaries where a merchant could safely find shelter and the distances between them. It is a document that has helped historians establish the lay of the Silk Road in the time of Augustus Caesar, that is, at the beginning of the Christian era.

On land the Silk Road stretched, roughly, from a point near modern Beijing to the eastern edge of the Mediterranean. (Broadly speaking, the term "Silk Road" can be stretched to include the sea routes used in later days, when there were too many bandits and wars along the land route.) In between lies some of the most dramatic and testing terrain the planet has to offer: the Gobi Desert, the Tien Shan range, and the Hindu Kush, as well as the steppes of Central Asia. If that sounds formidable, even today, consider what it must have been for a nomadic merchant leading a camel caravan loaded with goods.

The name "Silk Road" was coined by a nineteenth-century German explorer, Baron Ferdinand von Richthofen (a later von Richthofen was the famous Red Baron of World War I), who traveled through much of China and had some mountains named for him. Some anthropologists prefer to call the trading route the Scythian Way, a name that is not quite so catchy but is descriptive, for it honors some of the earliest traders along the route. Some have even suggested that the Silk Road was a path first made by *Homo erectus* when that ancestor of ours migrated from Africa to Asia.

The commerce along the Silk Road made fortunes, not just in cities such as Nishapur but in the oases and way stations all along its length and its branches. Their names today still stand for wealth and luxury and romance: Samarkand, Bukhara, Khotan, Xanadu.

Samarkand is still an important city in what we now call Uzbekistan, and when I heard that "The Treasures of Uzbekistan" were on exhibit at a gallery at the University of Pennsylvania, I went to see them. On display were merchant riches that hinted at the comfort and luxury the people of Samarkand had known at a time when our European ancestors were living in the squalor of the Dark Ages. The show featured beautifully glazed pottery, a trade good as well as a household necessity, and heavy silk clothing in rich magenta, reds, blues, and greens. There were tiles and frescoes from the mansions of Samarkand that had been made with great skill and artistry. But a more modest part of the exhibit brought one nomadic trader to life for me in a way that none of the citified luxuries could.

Nomads carried all their important personal possessions, along with their trade goods, through snowy mountain passes, desert heat, and bandit roadblocks. Nothing was more important to them than showing hospitality to wayfarers they met, however, and custom required that they serve tea to a stranger even before they asked his name. By lucky accident, the items needed for one merchant's hospitality had been preserved. He had carried his tea set in a protective case that could be attached to his saddle so that he would have it with him at all times. In shape it was very like the briefcases carried by modern commuters on the Long Island Railroad, complete with a strap handle. But the leather that covered it was densely embroidered with flowerlike geometric shapes, and the thread of the embroidery was, of course, silk, which still glowed in rainbow colors. Accompanying it was a fringed silk-embroidered-on-silk bowl cover to keep, I surmised, the flies from the sweetmeats that would have been served with the tea. A plain case and a plain lid would have served, but this merchant wanted the beauty and status that silk conferred.

There are scattered reports by archaeologists of silk appearing in the West at very early dates; one that made news a few years ago concerned an Egyptian mummy from 1000 B.C. that was said to have silk in its hair. This was held to be the earliest example of silk in the West and proof that the Chinese were trading silk at a much earlier date than had been suspected. But Irene Good, the silk archaeologist, is skeptical. The mummy's head, severed from the body, is on display in Prague. The silk may have gotten into the hair accidentally, and, although she has not had a chance to examine the silk, it might not even have been made by *B. mori.* Other early silks from the West that she has examined have turned out to be from other species of moths. Silk may have been smuggled out of China at an early date, but the Romans are not known to have had *B. mori* silk until sometime after 200 B.C.

The Romans were mad for silk, which was said to be worth its weight in gold and was associated with wealth, luxury, privilege, and a certain decadence. But people around the Mediterranean did not like the heavy brocades of the East, so women were set the task of pricking out the threads of imported silk fabrics and reweaving them into a light, diaphanous material. There was a good deal of huffing and puffing over this. Pliny the Elder in his *Natural History* wrote, when describing silk, "and so have the ends of the earth to be traversed: and all that a Roman dame may exhibit her charms in transparent gauze." Cleopatra's seduction of Mark Antony was rumored to have been the result of a silken garment, which allowed the full display of her breasts. In those times the Mediterranean world didn't have a very high opinion of women, so their taste for silk was regarded with a sense of what-can-be-expected-of-such-creatures, but to the consternation of commentators, men liked silk, too. All wearing of silk drew a frown from Seneca, the Roman philosopher who lived at the turn of the millennium. "I see clothes of silk," he wrote, "if clothes they can be called, affording protection neither to the body nor the modesty of the wearer, and which are purchased for enormous sums from an unknown people . . . Did philosophy ever teach men to wear silk?"

Tacitus once grumbled that "silk degrades a man," and in A.D. 14, Augustus forbade men to wear it. One of the complaints against the Huns when they overran the empire was that their tents were made of silk (by that time it was apparently forgotten that Julius Caesar had had canopies of silk). The great Christian schismatic Tertullian also considered silk unmanly; he wrote that when Alexander conquered the East, "all panting with the labor of warfare, he stripped his breast adorned with the emblems of armor, and covered it with transparent tissue; and was, as it were, softened, quenched, in floating silk."

When the Vandals seized North Africa as the empire crumbled, Procopius primly condemned the Romans as people who had grown soft because they wore clothes of silk. This is ironic, because Procopius was the gossipy but official journalist to the court of Justinian, emperor of the last shred of Roman Empire, Byzantium. And it was Justinian who managed, at long last, to steal the secrets of silk making from the East and break the Persians' grip on the silk trade. Here is the way Procopius tells that it happened:

> [In the year 552] certain monks, coming from India and learning that the Emperor Justinian entertained the desire that the Romans should no longer purchase their silk from the Persians, came before the emperor and promised so to settle the silk question that the Romans would no longer purchase this article from their enemies, the Persians, nor from any other nation; for they had, they said, spent a long time in the country situated north of the numerous nations of India — a country called Serinda — and there they had learned accurately by what means it was possible for silk to be produced in the land of the Romans. Whereupon the emperor made very diligent inquiries and asked them many questions to see whether their statements were true, and the monks explained to him that certain worms are the manufacturers of silk, nature being their teacher and compelling them to work continually. And while it was impossible to convey the worms thither alive, it was still practicable and altogether easy to convey their offspring. Now the offspring of these worms, they said, consisted of innumerable eggs from each one. And men bury these eggs, long after they are produced, in dung, and after thus heating them for a sufficient time, they bring forth the living creatures. After they had thus spoken, the emperor promised to reward them with large gifts and urged them to confirm their account in action. They then once more went to Serinda and brought back the eggs to Byzantium, and in the manner described caused them to be transformed into worms which they fed on the leaves of the mulberry; and thus they made possible from that time forth the production of silk in the land of the Romans.

Some authorities say that Procopius, who did his own editing and fact-checking, was confused — that the monks, who are reputed to have carried silkworm eggs in hollows bored into their staffs, actually got the eggs from the Persians, not the people of Serinda, if by that he meant China. There was some mystery, however, about where Serinda was and who the Seres were, and sometimes the words were used in a vague and floating way to mean almost any place or any people to the east or even those who traveled there.

At any rate, by the middle of the sixth century, *B. mori* had certainly made its way out of China, and in the East the secrets of silk making had spread by the third century to Japan and Korea. And another story has it that not much later a Chinese princess was married off to a Mongol prince and sent to live with him somewhere along the Silk Road. As part of her dowry she smuggled out some silkworm eggs in her hair. It might seem that no hairdo, even an elaborate one, could hide enough raw material to start a new industry, but a mere three-quarters of an ounce of silkworm eggs will hatch into 36,000 caterpillars if care is taken. Certainly it was not unusual for Chinese women to be married and sent off to cement useful foreign alliances. And since silkworm culture was traditionally women's work, it is not unlikely that it was women who spread the knowledge of the wonderful worms.

There are no equally colorful stories about how the silkworms came to Persia, but by Justinian's time (fifth century) a thriving silk industry there had its own caterpillars. And, once established in Byzantium under state monopoly, silkworm growing and silk making spread westward.* Turkey, Spain, Italy, and France all developed silk industries with their own silkworms over the following centuries.

In the sixteenth century, silkworm culture and a modest silk fabric industry were established in Lyons, France. In the following century the economic policies of Jean Baptiste Colbert, financial mender in the ruinous times of Louis XIV, helped Lyons overtake Italy. For the next several centuries Lyons was the silk center of the West, until diseases and genetic lack of vigor made silkworms difficult to raise there. At which point, as we know, our own USDA fell in love with silkworms.

Silkworms have always proved tricky to raise. There is no record that Europeans sacrificed prisoners of war on altars to the Silkworm Goddess, as had been the practice in early China, but a lot of rules and superstitions grew up around their culture. Women hung special pouches between their breasts to keep the silkworm eggs at the proper temperature for hatching. Loud noises were said to disturb the caterpillars, so the rooms in which they were fed had to be kept not only

---

* It is curious how often silkworm growing, the planting of mulberry trees, and the manufacture of silk were officially decreed by government and imposed on a country's population as not only an economic but a social policy. Governments seem to want to pull in a part of their national revenue by diverting some other country's foreign commerce, whether the proceeds go into a royal treasury or into the private pockets of the powerful. Moreover, the lower classes always look underemployed to those who rule them. In France, for instance, silkworm growing was encouraged by the government, and its purpose was expressed baldly enough by Henry IV, who observed that it employed "poor orphans and widows and the old" and kept money in France instead of allowing it to go to Italy. This approach held true for the Byzantine portion of the Roman Empire, England, France, the United States, and other countries as well. The Chinese, of course, had originated the practice.

warm and well ventilated but quiet. Children were forbidden to make loud noises or clap their hands while the caterpillars were growing. Thunderstorms came to be regarded as especially bad, for they were thought to attract evil spirits to the silkworms. It was possible, however, to prevent the spirits from entering the silkworms' delicate bodies by carrying live coals around the room. These had the added virtue, it was firmly believed, of soothing the caterpillars. Silkworms would thrive, it was said, if an object made of iron was placed in the room in which they fed. The room also had to be sweetly scented: the floor must daily be sprinkled with vinegar and strewn with lavender, rosemary, thyme, savory, and pennyroyal. When silkworms were done feeding, it was believed that they could be encouraged to spin by carrying fried onions around the room.

What brought the cultivation of silkworms to an end in Europe, as in the United States, however, was not possession by evil spirits, a lack of fried onions, or even the ills to which the finicky silkworms were subject. What ended it was the invention of rayon and other man-made fibers that had some of the same qualities of silk but were cheaper and easier to produce. Silk is still a luxury good, but today most of it is produced in the East.

China, the original home of *B. mori,* produces 90 percent of the world's silk, Marian Goldsmith told me when she took me into her laboratory to show me the silkworms she was growing. "It is still a cottage industry in the Orient," she said. "In Thailand it is almost exclusively so." The Chinese are doing interesting work on silkworm genetics, but it is not widely known, because few researchers in other countries can read Chinese, and the Chinese, in general, do not publish in English. So the Chinese may still know a thing or two about silk that the rest of us do not.

In her lab Marian pointed to a pile of finished cocoons, fluffy with silk. Some were golden in color, which marked them as a particular Chinese strain, and some were white, like the ones in my framed Kadonaga sculpture. Several batches of growing silkworms were feeding steadily in rectangular plastic dishes held above their own frass, or excrement, by a screen-wire grid. They were big, creamy white caterpillars with dark markings on their heads, which gave them a surprisingly appealing "face." They moved languidly from one green lump of artificial feed to another. Artificial feed frees researchers from the need to have a constant source of fresh mulberry leaves. It is expensive, probably too expensive to use in a cottage industry, but it does allow research to go on throughout the year in all climates. Marian said it was made up of a combination of rice bran, cellulose, mulberry powder, fishmeal, soymeal, and vitamins, plus nutritional elements that first attract the caterpillars to it and then make them bite into it.

A silkworm's ability to accept and thrive on an artificial diet is a matter of genetics. Japanese breeding studies in the 1970s ended up producing a strain of silkworms with a gene-driven taste for a great many foods: the leaves of cabbages, plums, cherries, and persimmons as well as the artificial diet. Over the centuries silkworm strains have also been developed to bring out another genetic trait: the ability to produce several generations a year rather than the single one produced by their presumed ancestor, *B. mandarina*. This is potentially a very profitable trait, but in the past it has been usable only in places where a constant supply of fresh mulberry leaves was available. But when combined with the mutations for a lack of choosiness in feeding, this trait is much more useful to industrial silkworm growers as well as to researchers.

Some of the mutations in silkworms were artificially induced in the laboratory. Geneticists have learned how to create mutations in an organism whose genome is somewhat familiar; in the process they can apply the understanding they gain to other organisms that are of wider economic interest. In the case of silkworms, the mutations that allow them to eat a wider variety of food also offer a genetic model for the study of moths whose caterpillars eat so many different kinds of plants that they are considered agricultural pests.

Marian picked up a silkworm and handed it to me. It was about as long as one of my fingers but plumper, and it looked, partly because of its facial markings, enormously pleased with itself. It felt cool, leathery, and helpless.* It moved lazily, checking out my palm for food ever so slightly. But Marian said that it was done feeding and nearly ready to pupate, to spin its silken wrap.

---

* The sixteenth-century poet Thomas Muffet, the author of the lines at the head of this chapter, also wrote of the care of "silkwormes" in his would-be georgic. He was a physician, farmer, and slavish admirer of Virgil.

> Forbeare . . . to touch them more than needes,
> Skarre children from them given to wantonnesse
> Let not the fruit of these your precious seedes,
> Die in their hands through too much carelessnesse:
> Who toss and roule and tumble them like weedes
> From leafe to leafe in busie idlenesse,
> No squatting them upon the floore or ground,
> No squashing out their bellies soft and round.

I asked Marian if there was any way this caterpillar could go feral, survive in the wild. "No," she said. "If it was feeding and fell off a leaf it wouldn't be able to crawl back up a tree. It would starve to death." She went on to explain that the parent moths were so heavy that their flight muscles were not strong enough to carry them. "It isn't that their muscles are atrophied, because when the males copulate they flutter their wings over the females, but they just can't lift up those big bodies they now have." This lack of flight ability would also make mating in the wild a doubtful matter and, combined with their lack of protective coloration, would make them easy pickings for a bird or other predator. *B. mori's* flightlessness, which makes it even more dependent on human care, is an exceedingly convenient trait for domestication. So the genetic modification for big-bodiedness, which we developed in them, serves an additional purpose beyond that of greater silk production.

Marian took back the silkworm I was holding and popped it into a brown paper bag with a crumpled paper towel in it. "This is what I learned to do when I last went to Japan to work," she said. "You read about all those complicated grids people build for them to pupate on, but they'll spin a cocoon perfectly well on a paper towel." She clipped the bag shut with a clothespin and set it alongside a row of similar bags. She opened one of them and, sure enough, the caterpillar inside had begun covering itself in gauzy wrap. When silkworms spin they secrete a gummy filament in one long continuous piece. The genes of the modern silkworm have been tailored to make a filament nearly a thousand yards in length. The filament is actually made up of double strands secreted by a pair of modified salivary glands in the caterpillar's head. As the double-stranded filament emerges from those glands the silkworm throws its head from side to side in a figure-eight pattern in order to envelop itself snugly and completely. Once covered, it begins to transform itself into an adult moth, but in the silk trade only the breeding stock are allowed that privilege. The rest are killed by heat as soon as they are done spinning so that they will not emerge as adults and break the filament, for then it would be impossible to reel.

Marian is not interested in reeling silk, so her moths are allowed to emerge, and she carefully mates them, a painstaking job. "I've got to go to Japan in a few months," she says, "but I can't be gone long because it is hard to train someone to baby-sit my silkworms, and if the animals die there's one whole line gone and I can't go on with my research."

The genome of *B. mori* is estimated to be made up of about 530 million base pairs, those rungs on the spiraled DNA ladder, those notes in the variation-upon-a-theme that is the genetic code of a species. In comparison, the fruit fly genome, which has now been mapped, is made up of 140 million base pairs. Fruit flies have given us an introductory understanding of basic genetic processes, but the larger size of the silkworm genome can add a layer of depth to that understanding.

In our five-thousand-year association with silkworms, we have conducted a great many genetic experiments, intended or not, upon them. Mutations have been preserved that would have been lost in the wild. A great body of knowledge about heritability in silkworms grew up long before human beings ever knew the word "gene," let alone "microbiology." And over that time significant races, with their own distinct allelic frequencies, have been bred. Those races, or strains, are comparable to the purebred strains of mice that commercial laboratories create for mammalian genetic research.

Those races of silkworms — Japanese, Korean, Indian, Thai, European, and others — have developed from handfuls of eggs tucked within a princess's hairdo, hidden in the hollow of a priest's staff, or sneaked out of China by other means centuries ago from silkworm populations whose breedings were already severely restricted and controlled by the creatures' own lack of mobility and need for human care.

Even to this day, silkworms from China are the most diverse genetically and feature a greater array of alleles than other silkworm populations do. Of course, no individual can have the alleles present in an entire population. Before silkworms became dependent on humans for mating, the alleles were shared around in a characteristic and constant ratio, as they are among animals in any wild population. But once the silkworm moths could no longer mate freely, distinct genetic lines would have begun to emerge.

And then, when a princess or a priest snatched up a handful of eggs and smuggled them away to establish a new population in a new place, it was only the handful of alleles that the handful of eggs possessed that could be inherited and spread through the new population. It would be physically impossible to recapture the home population's alleles in that group's same frequency, let alone in the frequency of the wild species. Therefore the silkworms in each new place became different — in looks, behavior, and genetics — from their Chinese cousins.

Biologists call this phenomenon the Founder Effect, the result of a small, isolated group of individuals imposing their own characteristic genetic profile on descendants. The Founder Effect is at work every time a population of plants or animals moves beyond its place of origin and can breed only within a small group. The Founder Effect is shown in the establishment of regional differences in blood types among isolated human populations, for instance. The blood types A, B, and O are controlled by three alleles of the same gene in human beings, and widely separated populations perpetuate the alleles of their founding ancestors. Pure Peruvian Indians, as one example, have 100 percent type O blood, while among African Pygmies, 31 percent are type O. Among American whites and blacks, whose ancestors traveled hither and yon and chose husbands and wives from a wide array of groups of people, some 45 percent are type O.

Because silkworms not only allowed but required controlled breeding, a breeder could easily select unusual alleles to become characteristic of a particular line. Various mutations could be cosseted, various attributes enhanced — the precise kind of filament required by those Japanese musical instrument makers could be produced by genetically modifying the opening of the spinneret from which it emerged. Strains could be created that would produce more silk or a longer filament or would be better suited for a particular climate, as well as those that could eat a variety of food or glow green.

In the wild, most genetic changes are tested against life's toughest question: Will they better suit the animal to survive and reproduce successfully? But survival in the wild is no longer pertinent when the animal is brought under our control. We pose other questions. Of the silkworms we ask: Can't you behave more quietly and be less twitchy? Can't you be bigger and heavier and make more silk? Can't you eat this nice artificial food? Silkworms must answer yes to those questions and others like them in order for us to allow their lines to continue.

So *B. mori* is a highly "engineered" animal, which makes it an ideal laboratory creature for geneticists. One artificially created sex-linked gene, for instance, does not harm the male *B. mori* carrying it but does kill all the female eggs produced when he mates with a female of another line. This, according to Vladimir A. Strunnikov, the geneticist who described it, could be useful in silk production, for males produce 20 percent more silk than do females. It is also helpful to geneticists who need a purely male line. Other silkworm mutations are useful in research because they can be compared to homologous genes in other animals in order to puzzle out basic biological processes.

I asked Marian what she thought of the flap over *Bt*-corn. No woman who has a fluorescent silkworm as an office pinup is likely to be shocked in the way the public seems to be. She said that researchers could investigate the whole problem of adding genes for toxins to which the target animal is developing resistance. And she saw it as a good research problem for silkworm geneticists because the corn borer is a moth, after all. The moths that survive *Bt* sprays have certain alleles of certain genes that allow them to live. They may pass those alleles along to offspring, and a new population can grow up that is not affected by *Bt*. That resistance takes many generations to develop, and the precise genetic mechanism is complicated and not well understood. If it could be worked out using silkworms as a model, strains of *Bt* (there are many) could be developed to which it would take longer for the corn borers to become resistant.

Japan has now funded the mapping of the silkworm genome, a map that will lead to greater understanding of many biological processes. The Japanese learned, somewhere around A.D. 200, the secrets of silkworm rearing and silk making. Just like people in the West, the Japanese liked and wanted silk and had offered a large bounty for the knowledge of its making. A group of Koreans collected the bounty by kidnapping four Chinese concubines and bringing them to Japan, where they were forced to tell everything they knew about silkworms. But this was long after trade had introduced Chinese silk to the Japanese. Japan became, in a sense, the eastern terminus of the Silk Road.

It was not too many centuries after Japan began its silk industry that another animal species minted by human ingenuity came from the West along the Silk Road and became an aid in the production of silk. This species, however, had a part in its own making. And its name was Cat.

Of Lions, Cats,
Shrinkage, and Rats

[In Egypt] all the inmates of a house where a cat
has died a natural death shave their eyebrows.
— Herodotus, *The Histories* (fifth century B.C.)

D URING THE 1930S, when I was a child in Kalama-
zoo, Michigan, a lion lived halfway between the
post office and the city hall. My father's office was
in that city hall, and when I was allowed, I would walk over
and look at the lion. The lion, a male, lived in the office of the
Chevrolet dealership, and I could see him through a big plate
glass window. He had a magnificently rough and shaggy
mane, but his coat looked a little patchy and his teeth were
worn. Occasionally he paced up and down, lashing his tail lazily, but more often he slept, sprawled across the top of a pair
of desks shoved together against the wall. What I liked best
were the very rare occasions when he would sit up, open his

OPPOSITE: Cat. Adapted from Edward Topsell, *The Historie of Four Footed
Beasts and Serpents and Insects,* 1658.

mouth, and mildly roar. When he did that he looked just like the Metro-Goldwyn-Mayer lion.

To this day I don't know why the lion was there. I never thought to ask at the time. Children accept whatever is as ordinary, and, in addition, I lived a life rich in what now seem to me to be unusual animals. My father was the city's park superintendent in those days, and that job included superintending the city's small zoo. The monkeys in the zoo couldn't overwinter in their outdoor quarters, so in the cold months they were moved to a sawdust-filled loft above the city's truck and snowplow garage. Sometimes my father had to go down there on a weekend morning to get a work crew started clearing snow from the ice-skating ponds, and he would take me with him. While he talked with the men below, I would climb up the wooden ladder into the warm loft and watch the monkeys. That didn't seem unusual to me at the time, nor did bottle-feeding the baby black bear in the shed beside our garage. The baby bear had been delivered to the zoo while it was still too young to be put in with the other bears, so my father had brought it home and put it in the shed until it grew a little bigger. I can remember watching my older brother feeding it from a nursing bottle that I had outgrown only a few years before.

Today it strikes me as odd — and sad — that the lion lived out his life in a Chevrolet garage. I can't imagine that the dealer who kept him there got much satisfaction out of the arrangement either. Lions and the other big cats do not tame nicely. Biologically speaking, they are not house cats. But then, at one time, house cats weren't house cats either.

As I write, Black Edith Kitty is sprawled across my desk on

top of several important stacks of paper I need. His posture is much the same as that of the Chevy garage lion, but fortunately he is much smaller. I can easily lift him up and pull out those papers when I really, really need them without any fear except that of disturbing a contented animal. He comes from a line of ancestors genetically modified for human companionship.

Those ancestors are much younger in biological age than the ancestors of the silkworm cocoons in their frame on the shelf behind my desk. Insects had their beginnings somewhere around four hundred million years ago, while mammals began some two hundred million years later. The first mammals were little scuttling things, nothing like either the Chevy lion or Black Edith. The tree-living animal that paleozoologists consider the first catlike carnivore, ancestor to both kitty and lion, appears in the fossil record only thirty million years ago, and distinct modern cat ancestors separated from the proto-cats less than seven million years ago. Then, during the Pleistocene era, which began a mere two million years back, something curious happened to a lot of mammals. (Relatively modern humans also appeared in that era.) It was a time of unstable weather. Ice sheets grew and retreated. It must have been a hard time for mammals, because a number of them were completely extinguished and many of the remaining ones became smaller. Both things happened to the cats, the big ones as well as the small. Today we think of lions and tigers as big cats, but their ancestors were even larger. And the smaller ones, particularly those in the genus *Felis,* became even smaller. No one knows for sure why this came about, whether it was because of stresses from

The North African wildcat,
Felis silvestris libyca.

erratic climate or from humans with their fires, hunting, and increasing domination of habitat, or from some as yet unknown factor. But fossil evidence clearly shows over time a series of ever smaller wildcats, the ancestors of Black Edith Kitty. He and his kin in other households, barns, and alleyways are called by scientists *Felis catus.*

The considered opinion is that *F. catus* is descended from one race of *F. silvestris,* the wildcat. Both have the same number of chromosomes and a similar genetic profile. *F. catus* may also have a gene or two from a couple of other small species in the same genus.

F. silvestris is the wildcat that today inhabits, in declining numbers, Europe, Africa, and parts of the Middle East. A striped cat, a fierce and skilled hunter, it looks rather like a bigger version of our domestic tabbies. The northern race, *F. sylvestris sylvestris,* is the biggest of the species, heavy-bodied and chunky-looking compared to our pets and to the southern race. That race, *F. silvestris libyca,* is a smaller cat all around, lighter in build, high-chested, and long-legged. Wildcats and domestic cats seldom have the opportunity to mate, for the wildcat populations have been severely thinned through sport hunting and the disappearance of the wild places where they lived. Many (but not all) authorities say that the two cats are now so genetically distinct that such a union would be sterile. But over the centuries since they became separate species they may have exchanged a few genes.

The *libyca* race of *F. silvestris* separated from its European relatives about twenty thousand years ago, but there is little evidence that people took much notice of this elusive animal that stalked its prey in dim light. As with corn or silkworms, there is no neatly recorded account of who took the first step toward domesticating cats, who realized that they might be pleasant animals to have around, useful as predators upon small rodents and handsome to look upon. But there is evidence to suggest that about ten thousand years ago in Africa, the *libyca* race of cats and the Egyptians, who fancied animals in general, began keeping company.

By that time agriculture was well established. Agriculture meant the harvest of grain, which needed to be stored. Stored grain attracted mice and the small wild rats of the countryside. Anyone who has ever lived with *F. catus* will probably see the good sense in the considered zoological opinion that it was the cat who came to man, not the man who gathered in the cat, taught it, and disciplined it into being a good mousing machine. Cats are not amenable to instruction, but they *are* good at finding food, and because

F. s. libyca was a small cat (though bigger than Black Edith and his kind), it could not bring down anything like a gazelle: mice and rats were its proper-sized prey. Probably the cats were simply tolerated around granaries for a long time; evidence of true domestication appears only a few millennia ago, about the time Empress Hsi-Ling Chi was dropping the cocoon into her tea in the palace garden in China. But at some point Egyptians took cats into their houses, began to control their breeding and reproduction, and gave them a name: *miaw.* (The word "cat" comes from the language spoken by Berbers living to the west of Egypt.)

It is intellectually suspect to seize upon an attitude found among a conservative contemporary people and use it as an example of what people must have thought thousands of years ago, but with that caveat in mind, it is interesting that present-day Berbers regard cats as special in the same way that, evidence hints, the Egyptians did. Lloyd Cabot Briggs spent considerable time with Berber nomads and in 1967 wrote:

> Cats are to be found in almost all Saharan settlements . . . They are valued mainly for their enthusiasm and skill in killing scorpions and poisonous snakes, but they are rarely given any special care. The males are occasionally eaten ritually . . . and they are sometimes bred and fattened expressly for this purpose. Sedentary women eat cat meat as a medicinal specific, or to promote general well-being or fertility, and the charm is thought to work best if the pieces of cut up meat are swallowed whole without being chewed. Each spring the women . . . are led by the chief woman of the town up out of the valley to the plateau above, and there they have a general jollification and ritual picnic with

cat as the main course. Throughout North Africa cats seem to inspire some sort of magico-religious awe, but apparently this question has never been investigated systematically.

There is a good record of the high regard the Egyptians had for cats, however. They not only painted and carved statues of them but mummified them as well — mummified great numbers of them, for they were, or at least became, god-things. One Egyptian deity has the head of a cat, and at least some of the mummies found in vast cat cemeteries appear to have been votive offerings; they were killed young — under a year in age — and may even have been raised for sacrifice. Occasionally demand must have exceeded supply, for a few mummies contained the bones of the marsh cat, a larger animal of a different species. The mummies also show that the cat-wrappers may not have been above a little bait and switch. Some of the "cat" mummies when unwrapped prove to be an assortment of odd animal bones and other materials.

Throughout the Mediterranean world there was a principle designated as deity that had to do with the fertility and productiveness of plants and animals — wild animals as well as livestock. This deity, female in aspect, was in general benign but needed to be won over by sacrifice when she became temperamental. She was helpful both in childbirth, the emergence from the dark and unknown, and in death, the disappearance back into the dark and unknown. Different cultures gave her different names. In death's aspect she was sometimes known as Styx, and from her the underworld river took its name. In Greece she was Artemis or Hekate (Persephone had some of her qualities also). In Rome she was Diana or Trivia. Throughout the region she was a hunter, and the name Hekate comes from a Greek word that means "far-darter." Her hunting abilities combined with her helpfulness in agriculture made her the protector of grain. Ovid wrote slyly of the epic battle between the gods and the giants, which, contrary to the conventional accounts, the gods *lost*. In their shame they fled to far parts of the world in animal disguise. In Ovid's version Diana-Trivia, the hunter, went to Egypt and became a cat.

Bastet was the name of this cat-goddess in Egypt, and she had been there long before Ovid's day. She was depicted as standing tall with the body of a woman and the head of a cat. In her destructive form she was a lion, but when placated she was a cat.

For the Egyptians, a household cat was a pet with a hint of the divine, and it would not have seemed at all odd to raise a cat for a talisman and as a sacrifice, a votive offering preserved by ritual mummification.

So many cats were mummified that toward the end of the nineteenth century boatloads of excavated mummies were exported from Egypt to England as ballast on the home voyage of commercial ships. They were ground into fertilizer after they reached port. One nineteen-ton consignment thus disposed of was estimated to have contained eighty thousand mummified cats. Some, however, were saved from the fertilizer factory, and it is from them that we can see the beginnings of the genetic change brought about by domestication. It is a story of diminishing size.

Gradually, beginning sometime after 2000 B.C. and continuing on through the centuries, a curious thing happened to

Bastet, the cat-goddess of Egypt.

cats, a curious thing that has been repeated every time a large wild animal, be it cat or cow, has taken our fancy. If you decide to bring wildcats into your house and make them your companions, you'll pick those that appear to be less frightened of human beings than the ones that run away, and you will take the smaller ones, which will be easier to control and less likely to hurt the children. You will take the more beautiful ones, too, those with special markings.

All of these qualities — size, fearfulness, markings — are controlled by various genes or combinations of genes. Just as with silkworms, the Founder Effect began to work on cats. No one cat, be it *F. s. libyca* or *F. catus,* carries all the alleles of the genes that can be found in the entire population of its kind, merely a fraction of them. The cats that the Egyptians took into their homes expressed certain alleles that made them smaller than average, prettier, and less easily spooked. Some of the domesticated cats would have escaped in the lustiness of breeding that is characteristic of cats, wild or domestic, and the genes of those would have mixed with the genes of their wild relatives once again. But many would have mated with other house cats who also possessed the more admired genetic traits — a particular color pattern, a particular friendliness, docility, and daintiness of size. In each generation animals showing the genetic traits that humans liked would be bred with one another, and those that expressed less desirable traits would not be bred. (Please! You've got to get rid of that great big vicious *miaw!*) By the middle of the second millennium B.C., *F. catus* had become genetically distinct in important ways from *F. s. libyca.*

In the wild those traits that are adaptive for survival and reproductive advantage are brought out through natural selection. So cats that were fierce, furtive hunters, alert to the snapping of every twig, with coats that gave them good camouflage, would have been favored by evolution. But again, as with corn or silkworms — or dogs or pigs or cattle, for that matter — the traits that suit a cat for a wild life are not necessarily emphasized in artificial selection by humans.

Artificial selection can transform an animal very quickly,

as one study has shown in a rather dramatic way. On farms where foxes are raised for their pelts, it is convenient to have animals that are tamer than the wild variety, so a Russian experiment attempted to selectively breed foxes for tameness. Foxes that were more tolerant of being handled by humans and fed by hand were interbred, and the friendlier offspring were selected as parents of the following generation. And so on. To the researchers' surprise, it took only fifteen generations to create foxes who came when they were called, liked being petted, barked like dogs, and wagged their tails when they saw humans. Although the breeders had selected only for genetic traits that affected behavior, an unforeseen result took place: the friendly foxes also looked different. Some had coats patched with white spots, others had floppy ears. Some began to develop rather doglike faces, more juvenile in character, with shortened noses. When they were injected with a stress-producing hormone, the tame foxes showed less of an adrenaline response than the wild control animals did. Through the generations, the hormone-producing ability of their adrenal glands had grown weaker. This, it turns out, is not unrelated to the changes in the foxes' skulls and is similar to what happened to cats long ago in Egypt.

Helmut Hemmer, a German scientist, has found that certain genetic modifications are necessary to produce biologically calm animals suitable for complete domestication, whatever the species: wolf to dog, wild pig to barnyard variety, or any other animal, including wildcat to lap cat.

Hemmer discovered, from archaeological evidence, that not only are domestic cats descended from the smallest-bodied race of wildcats of their time, but also, like dogs, pigs, donkeys, and many other transformed animals, they shrank even more as they became more closely associated with us. In addition, the skulls of mummified cats show that, even allowing for their smaller body size, the cats' brains shrank in a *disproportionally* rapid way over time. The shrinking of the skull forced a rearrangement of jaw structure and led to a corresponding rearrangement of teeth in the emerging new species. Casts made of the insides of the empty mummified skulls still carry the imprint of the brain's pattern upon them and show that the neocortex of domestic cats became smoother and less folded over time than those of their wild cousins. And the brain centers for sensitivity to changes in sound and movement are reduced in domestic cats, compared to their wild ancestors. They are less jumpy and can accept the fact that sometimes twigs snap. No bother.

The brains inside the skulls of the Egyptian mummified cats were about the same size as those of the smallest of their wild ancestors. By the Middle Ages, the European cat brain was 10 percent smaller still, the same as the average cat brain of today — except for Siamese cats, whose brains are yet another 5 to 10 percent smaller.

Robert Williams is a neurobiologist who studies the evolution of the brain. He acquired two adults and one fetal kitten of the Spanish wildcat, *F. s. tartessia,* which have changed little from wildcats of Pleistocene times. They have lived in a warm, isolated environment and have, so it is believed, been less subject to evolutionary stress than either the northern wildcat, *F. s. silvestris,* or the African one, *F. s. libyca.* Assuming that his three wildcats were nearer to the ancestral stock than any others, Williams compared their brains with those of unusually large domestic cats that were nearly comparable in size. He found that in the part of the brain devoted to vision, the domestic cats had 30 to 35 percent fewer neurons than the Spanish wildcats.

The shrinkage of the cat's neocortex during domestication was matched by the shrinkage in size and activity of their adrenal glands. The adrenals lie near the kidneys and secrete a variety of hormones, among them those that we call the "fight-or-flight" hormones, adrenaline and noradrenaline, which influence an animal's quickness of response to perceived danger.

My brother once brought a raccoon kitten into the house to live with us. He was filled with the optimistic belief that he could tame it and that it would be quite like a pet cat. The baby raccoon was cute and playful for a time, but it soon grew up and turned into a wild animal. A full complement of adrenal hormones must have pumped through its body when one of the dogs wandered into the room it was in, when the telephone rang, when someone walked quickly down the hall to answer a knock on the door. It ate the goldfish out of their bowl and neatly scooped a pet turtle out of its shell and ate it, too. Curtains dangled limp and shredded, the result of too many frantic climbs to perceived safety from perceived danger.

Even a small wildcat, with a full stock of vision neurons and actively functioning adrenals, would have behaved like the feline equivalent of that raccoon. The wildcat would startle easily, arch its back, spit, claw, and dash up whatever ancient equivalent of curtains was nearby. But a representative F. catus would not. Body size, brain compositional pattern, and adrenal gland function are all genetically determined. Controlled breeding can bring out the alleles of the genes that influence these physical structures and thereby bring about changes in behavior. We have genetically designed pet cats to be mellower than their wild relatives. (Of course, that doesn't mean we can't turn them into spitting, hateful neurotics when we mistreat them. They can learn to be distrustful and wary.) And we did it long before we knew the word "genetic" or could pair it with "engineering."

Hemmer calls this mellowing process "the decline of environmental appreciation" (which is the title of his book). In other words, dogs and cats are not as jumpy as wolves and wildcats because they don't notice as much. Predators need to be alert and responsive to the least sound, the slightest movement. An equally responsive household pet would be a nuisance.

Just because the cats we share our homes with have been genetically modified to be smaller brained and less reactive, it does not follow that they are any less "intelligent" than their wild cousins. Their association with us has given them other things to think about. Anyone who has ever lived with a cat has smart-kitty stories to tell. And, of course, I have mine.

Wildcats are strongly territorial. But along with altering the domestic cat's size and reactivity to make it easier to live with, we have also rejiggered its territorial sense enough to allow us to take it from place to place. *F. catus,* the descendant of *F. s. libyca,* boarded ships bound for new continents, even traveled the Silk Road. Black Edith Kitty was born on a farm in Missouri. Nevertheless, he set up a fresh territory without visible bother at our new home in Maine; he also takes curiously well to regular two-day car trips from Missouri or Maine to Washington, D.C., where my husband's house is. During those trips he snoozes. Coming from the west (Missouri) he wakes up at the Kennedy Center; from the north (Maine), at the Chevy Chase traffic circle. In either case, his behavior is the same: he begins to pace the dashboard and say his name in ancient Egyptian, and he does it long before the dogs in the car show any awareness of where we are. Both places are several miles from the house, and I don't believe that the cat has ever sneaked out to the opera or gone shopping in the commercial center near Chevy Chase Circle. Neither urban place has anything in common with the quiet, leafy neighborhood around my husband's house. When we pull up to the house, Black Edith bounds out of the car, runs up to my office, and sits elegantly on the bookshelf, waiting for the food and water bowl he expects me to unpack and fill before I do anything else (and of course I do).

F. s. libyca, with all its richness of neurons and all its wild talents, has a territory of its own making in a fixed place where generations of ancestors have also lived. Black Edith has a three-lobed territory that he has accepted from a human's choosing. He has accepted it with equanimity and skill. The territory is connected by corridors consisting of long boring rides inside an automobile, with an overnight stay in a motel, none of which make cat sense. His kind of intelligence, however, allows him to tolerate a state of affairs that would make a wildcat neurotic.

And then there is the matter of Black Edith's relationship to me. We have been together a long time, nearly twenty years, and throughout that time he has studied me with such close attention that he knows me better than any human does (including family members). He has studied me with some skewed but specialized part of his intelligence that his wild ancestors devoted to honing their hunting skills. And perhaps also their parenting skills. I have no secrets from Black Edith. His basic attitude toward me is that of a father mildly annoyed at my refusal to apply myself. Take mice, for instance. For all those years he has tried to teach me to take a proper interest in catching mice. Over time he has brought me many mice and, when possible, thrust them near my mouth so I can see how good they are and then learn to hunt them on my own. Not many months ago I awoke one morning to find a freshly killed mouse lying on the coverlet a few inches from my face. Black Edith was sitting alertly nearby on the bed, watching closely. He was full of hope. Once again, I simply didn't get the point and threw a perfectly good dead mouse out the window with an inappropriate comment. It must be disappointing for a cat to have a retarded human. I doubt that he would have been so patient with a kitten of his own. I am humbled by his affection. A wildcat would simply have run away from me long ago, having no interest in attempting to nourish or evaluate another animal so different and so potentially dangerous and so given to making meaningless journeys.

All of the changes in cats through the process of domestication — reduction in size, smoothing of the brain, depressed secretions of the adrenals, pruning of neurons — bred in certain characteristics and bred out others. Disregarding the role of mutation here for the moment, there is evidence that a lot of the alleles that made many of these

changes possible were already present in *F. s. libyca* populations. Recent studies of allozyme variation have demonstrated this. Allozymes are variant forms of enzymes, those
catalysts of life's biological processes. And since their production is under genetic control and they are heritable,
studying allozymes is a way to study genetic change. It is one
of the ways, for instance, that the *F. catus* lineage has been
tracked back to *F. s. libyca* rather than to some other race of *F.
silvestris,* for the two have greater allozyme similarities than
others. But studies also show that all of *F. silvestris* has much
more built-in allozyme variability than other cat species do
(the clouded leopard turns out to be the least variable, most
genetically stable cat). In other words, the stock that produced our pet cats was easier to change because it had more
genetic possibilities for change, and it continues to be easier
to change, as shown by the ever-fancier varieties that cat
breeders bring forth.

Mutations — the alteration of DNA bases in brand-new
ways — also produce change, and if those changes don't kill
the animal at an early age they may be inherited. Most mutations turn out to harm the animal or to make no appreciable
difference. But a very few can make the animal more fit in
the genetic sense (for domestic animals that may mean that
we prefer the mutation). These mutations will be passed on
to future generations, spreading through the populations, sta-
bilizing eventually at a certain predictable ratio. When mutant animals are taken to a new place where they serve as a
founding population, their descendants will have a higher
frequency of the mutation than the original population did.
By studying and comparing the ratios of these frequencies,

population geneticists can make reasonably good judgments about the spread of animals into a new place. Neil B. Todd has done just that with cats.

Cats make a splendid study in this respect because they feature several mutations that are expressed phenotypically — that is, they show up for us to see — including mutations that produce different coat colors and numbers of toes.

The genetic process for determining coat color is particularly interesting in cats, for people have made questionable links between behavior and coat color. In mammals there is an intriguing and not well understood connection between certain endocrine hormones and the black-brown pigment melanin, which is produced by an enzymatic process. The hormones and the pigment share a metabolic pathway, which means that they are produced by the same chain of biochemical events. The typical coat color of *F. s. libyca* appears to be a plain, pleasant, tawny orangish brown marked with deeper stripes of brown that grow ever darker down the tail, which is, at its tip, sable black. On close examination, however, it can be seen that every hair is lightly banded in color. A coat of this sort allows a light-footed, stealthy small predator to virtually disappear into tall grass shadows or dappled shade. Any mutation that results in a less well camouflaged coat would put a wildcat at a disadvantage and would therefore tend to disappear from the population.

This stripe-haired, shade-and-shadow coat pattern, which is typical of many wild animals, is chiefly controlled by a gene that zoologists have named *agouti*, after the neotropical rodent with a coat of that sort. But now and again animals of many kinds are born with a mutant gene called (rather unimaginatively) *non-agouti,* which gives them coats made up of unbanded, solid-colored hairs. *Non-agouti* animals do not look brindled and stripy, shade and shadow.

A Georgia State Hospital researcher, Clyde Keeler, years ago studied the relationship between coat color and behavior in caged rats, mice, mink, and foxes. He concluded that the animals with solid-colored coats were less likely to run away from their handlers and act frightened than were those with the *agouti* coats. Among foxes, the ones with silver-blue coats were the least fearful; the chocolate brown ones were a very close second. Among the other animals, those with the darkest brown or black solid-colored coats were the least fearful. Keeler concluded that because production of the pigment melanin and the hormone adrenaline shared the same biochemical pathway, the two characteristics "may conceivably" be linked.

Unfortunately, Keeler's tentative and qualified linkage between dark coat color and calm behavior in rats, foxes, and mink lost its tentativeness and qualification as it was cited over the years and has been extended, without additional research, to cats as well. Later scientific papers footnote Keeler's work to support the assertion that cats with dark coat colors are less fearful and nervous than ones with lighter coats, and popular books simply state that black cats are mellower than cats of other colors (often with the observation that, as everyone knows, redheaded human beings are more volatile than anyone else!!!).

Even if Keeler's conclusions are valid (and a later Russian study on foxes clashes with his in some respects), it is problematic to generalize to animals other than the ones he experimented with. His rats, mink, and foxes were wild animals and were kept captive inside fences or cages, a situation that would have skewed their natural behavior right from the beginning. In addition, he judged their fearfulness by measuring how close a human being could come to them without the animal running away. It is hard to see how a similar test, even if it could be made, would apply to domestic or even feral cats, whose behavior has been modified by the learned relationship every cat has with humans and the accommodation that generations of its ancestors have made with the human environment.

Neil Todd's study, however, used coat-color mutations in cats, as well as those that grant them extra toes, to trace the animal's historic spread from their native Egypt. By statistically charting the current prevalence of cats with certain coat colors and numbers of toes in various places, Todd was able to make "cline maps" showing the graded variation of coat colors and extra toes around the world. The maps turned out to reflect historical trade routes. Cats with extra toes, such as the Maine coon cat, are frequently found in eastern New England, and his statistical analysis of their frequency in different places suggested that the mutation had been present in cats around the port of Boston by the middle of the eighteenth century and had then moved on up the coast to Halifax, Nova Scotia, following the shipping route.

The mutation that gives male cats a ginger-colored coat and females ginger, tortoiseshell, or calico coats produced a particularly telling map. The orange mutant gene is found only on the X, or female, chromosome. As with humans, female cats have paired sex chromosomes, XX, and male cats have XY sex chromosomes. The female cat, therefore, can have the orange mutant gene on one X chromosome and the genes for a black or white coat on the other, and those can affect or modify the orange mutant gene. If that is the case, those several genes will be expressed in a blotchy coat of the tortoiseshell or calico kind. But the male, with his single X chromosome, has only one of that particular coat-color gene: he can be not-ginger or he can be ginger (although some modifier genes can add a bit of white here and there), but unless he has a chromosomal abnormality he cannot be a calico cat.

In Europe the proportion of contemporary cats bearing this orange mutant gene shows a distinct concentration along a line running from the Mediterranean port cities of France and Italy, where Todd surmised the mutants traveled from Egypt, through Europe to Britain. The highest concentrations followed the valleys of the Seine and Rhone, the shipping route between north and south, augmented by a system of canals.

Other color mutants also showed a striking correlation with paths of commerce and migration. What Todd, the population geneticist, had worked out mirrors what archaeologists and historians have learned about the way domestic cats spread from North Africa.

Water can be a barrier to the dispersal of many animals, but just as Black Edith Kitty takes readily to commuting a thousand miles by automobile, so, too, cats more than two thousand years ago began to travel, first by ship and later by caravan. Since the cats of Egypt had a touch of the sacred, they may have been carried on shipboard as talismans for a safe voyage. But cats are drawn to the exploration of nooks and crannies, and some may have simply wandered onto ships out of curiosity and become unintentional stowaways. Black Edith has several times boarded an open UPS truck parked in front of the house, and Ted, the delivery man, has told me that he often has to deliver hitchhiking cats back home. It would not have taken a ship's crew long to discover that in addition to being talismans, cats were very clever at killing the mice and rats that chewed on the trade goods in the hold. They were also beautiful to look at and fun to have around, considerations that often are more important than any others when a human being makes a decision.

Archaeological and pictorial evidence tells us that by 900 B.C. cats had found their way to the Mediterranean island of Crete, close to Egypt, and soon after were known in mainland Greece. By 300 B.C. they were in India, and one hundred years later in China, where they are mentioned in the court records of the early Han dynasty.

Those particular dates are significant, because they coincide with the beginning of the regular, if initially illegal, trade of silk from the East to the West. And cats, unusual enough to be owned by royalty, were, according to mercantile records, among the trade goods taken east as payment for silk.

There are other species of smallish wildcats of the genus *Felis* in the East, and some of them look so like certain modern breeds of cats, such as Siamese or Persian, that in the past those species were often cited as their ancestors. But detailed studies of the wildcats' chromosomes, allozymes, skulls, and skeletons have concluded that they probably are not, although one eastern race of *F. silvestris,* called *F. s. ornata* or, more familiarly, the Indian desert cat, a spotty animal, did very likely contribute to the making of *F. catus.*

In the East cats came to be closely associated with Buddhism, for temple cats guarded manuscripts against mice. That, apparently, is the reason they at last made it to Japan, for their arrival there coincided with the arrival of Buddhism in the sixth century A.D. Legend has it that a few centuries later a Japanese emperor was so struck by the birth of five pure white kittens born to a white mother cat imported from China that he decreed they should have the upbringing of princes. These were apparently not albinos (whose lack of color is caused by a double dose of a mutant recessive allele) but, because they were said to have blue eyes, probably very early examples of the cat we now call Siamese. The Siamese coat color is caused by a complex interplay of alleles of the albino gene with an allele of another gene named *chinchilla.* The resulting coat is unusual because it is sensitive to temperature: warmth keeps the coat light and cold darkens it. Kittens born to a Siamese mother emerge from the warmth of the womb pure white. As they grow, their coats become creamier and become darker at the ends of the tail, ears, muzzle, and paws, where their body temperature is lower. If, through mischance, they lose a part of their coat, the hair that grows back is darker, too. Siamese are also frequently mutant in another way. Like the cats along the northeastern seaboard of the United States, they often have extra toes.

The emperors' cats were so privileged that cats came to be cherished widely in Japan. They were allowed out-of-doors only on leashes, which enabled their owners to control their breeding. And they were honored, even revered, for their service as mouse guards not only in Buddhist temples but also in buildings where silkworms were raised. Mice had become a considerable pest of silkworms and their cocoons, and cats were held to be almost magical in their ability to make mice disappear. Before long it came to be believed that even a painting of a cat on the door of a silkworm establishment, or a statue of a cat within it, could scare away mice and rats. Artists were commissioned to make statues of cats in bronze, wood, and porcelain for premises lacking cats.

Not surprisingly, the images had no effect on mice, and because of this, public opinion, rather unfairly, turned against cats. They even came to be regarded as demons. The imperial court for a time tried to protect them, but in 1603 it gave up and decreed that all cats were to be turned out-of-doors to deal with mice directly. Buying, selling, or even receiving a cat as a gift became a crime.

Romans had gone to Egypt first to trade and then, under Julius Caesar, to conquer and farm. Egypt had cats; Rome did not. The Romans did not immediately show an interest in cats. They had traditionally used ferrets to keep down rats and mice, and some of their initial experiences with Egyptian cats were unpleasant. During Caesar's invasion a Roman soldier made the mistake of killing a cat. The Egyptians of the town rose up, lynched the soldier, and dragged his corpse through the streets. Egyptian resistance to Rome, according to the French cat historian Fernand Mery, coalesced around this cat indignity, and it was many years before the nation of cat lovers was decisively conquered.

It was not until the imperial period, which brought an end

to the turmoil following the murder of Julius Caesar, that the Romans thought well enough of cats to take them into their homes. As Rome extended its empire, cats extended theirs throughout Europe, traveling with the troops and administrators of the new provinces. Cats were welcomed in Britain, and archaeological digs show their presence in houses there after the Roman Empire collapsed, even as their wild cousin, the northern, heavy-bodied race *F. s. silvestris,* was being hunted nearly to extinction. The remains of a cat in an English cellar, trapped by fire raging in a wealthy man's house above it, shows that cats had spread that far by the fourth century.

That the cat was valued in Britain during the Middle Ages is demonstrated by a set of tenth-century Welsh laws setting the price of a cat higher than that for a pig, lamb, or goose and equal to that for a "housedog," though somewhat lower than the price for a hunting hound. And cats were treated tenderly, at least in some households, as shown in this fourteenth-century verse by Chaucer:

> Lat take the cat, and fostre him wel with milk,
> And tendre flesh, and make his couche of silk,
> And lat him see a mous go by the wal;
> Anon he weyveth milk, and flesh, and al,
> And every deyntee that is in the hous,
> Swich appetyt hath he to ete a mous.

There are many other affectionate written references to cats during the Middle Ages, not only acknowledging their useful role in keeping down mice and rats but also saying how companionable and beautiful they were thought to be, at least in England. On the European continent, however, there is less evidence that they were held in such high regard.

A century after Chaucer, Konrad von Gesner, a Swiss naturalist, wrote of the terrible effects cats could have on people. In 1658 his Latin writings on animals were translated into English by Edward Topsell, whose drawing of a cat, in a work entitled *The Historie of Four Footed Beasts and Serpents and Insects,* is the source for the illustration at the beginning of this chapter.

> It is most certain, that the breath and favor of Cats consume the radical humour and destroy the lungs, and therefore they which keep their Cats with them in their beds have the air corrupted, and fall into fever, Hecticks and Consumptions. There was a certain company of Munks much given to nourish and play with Cats, where by they were so infected, that within a short space none of them were able either to say, read, pray, or sing, in all the Monastery; and therefore also they are dangerous in times of Pestilence, for they are not only apt to bring home venemous [sic] infection, but to poison a man with very looking upon him; wherefore there is in some men a natural dislike, and abhorring of Cats, their natures being so composed, that not only when they see them, but being near them and unseen, and hid of purpose, they fall into passions, frettings, sweating, pulling off their hats, and trembling fearfully, as I have known many in Germany; the reason whereof is, because the constellation which threaten their bodies which is peculiar to every man, worketh by the presence and offense of these creatures: and therefore they have cryed out to take away the Cats.

During this period in Europe cats were eaten in stews, skinned for their fur, and immured in buildings during construction to scare off rats. When ancient buildings are demolished in modern times it is not unusual to find a dried corpse of a cat, sometimes accompanied instructively by an equally dried mouse or rat. This is true in England, too, where the obvious affection toward cats was blended with what we would today call cruelty.

In 936, even before Chaucer's time, the Welsh prince Howell Dda, who established the fair price for a cat, had issued a law for their protection and set penalties for "those who endangered its life, wounded it, or did not care for it properly." But for centuries after that law was issued — and who can say how well it was obeyed? — cats throughout Europe, and certain women who kept them, needed protection, which they seldom got.

There are many stories about cats being burned in fires, caged or tethered, singly or in numbers. The practice may have been a low form of entertainment for louts, drawing on the human capacity to take pleasure in inflicting slow and painful death upon small animals. But there is more to it than that. After all, the Egyptians, with their beloved household cats, also raised cats specifically for sacrifice, and their goddess, Bastet, required the burning of cats in her votive fires. (Bastet's counterparts, Artemis, Hekate, and Diana, all savored burned animals. At the time of the gods in ancient Greece, where domestic cats were not yet present, the sacrificial animal of choice for this deity was a black puppy. Black animals were often specified for sacrifices to propitiate the gods.)

Throughout human history new religions succeed old ones, but there is seldom a clean break between the two. The new attempts to suppress the old, but remnants persist —

driven underground, perhaps, or absorbed and transformed into the new. And this was true when Christianity replaced paganism. In the ancient world, Artemis/Hekate/Diana/Trivia/Bastet had been a benign but powerful figure who had to be placated. She aided in life's beginnings and endings, mysterious times of dark places. The moon and nighttime were hers, so cats were her perfect companions. Cats hunt in the dark. Their eyes, which seem almost to glow when seen in the light at night, were believed to change according to the phases of the moon. And black cats, in particular, wearing coats of the preferred color for sacrifice, could fade into the night, into the shadows and — who knows? — perhaps change into the shape of a woman. At any rate, Hekate, to whom Zeus gave so much,★ lost her benign aspects in Christianity and retained the frightening ones. She became Hecate, the enemy of God, the hag, the witch who did evil in the dark hours, accompanied only by her cat. Cats and women accused of being witches were buried alive, drowned (sometimes together in a big sack so that the terrified cat might claw and bite at the woman when both were submerged), or burned at the stake. Cats came to be regarded not as helpful and beautiful companions with a touch of the divine, as they had been in the past, but as harmful animals, the alter egos of witches, with a touch of the devil. A proper ending for both was death by fire. Folklore had it that a cat's tail and head,

---

★ The poet Hesiod, in the eighth century B.C., had written that Zeus honored Hekate more than the other gods:

> for he gave her gifts that were glorious
> to have a part of the earth as hers, and a part of the barren
> sea, and she, with a place also in the starry heaven,
> is thus exalted exceedingly even among the immortals.

thrown into the fire in the hearth, was enough to keep bad luck away from a household.

According to Neil Todd, the mutation that produces black coats in cats, an expression of the *non-agouti* gene, is a fairly recent one, having arisen probably no more than 2,500 years ago in the eastern Mediterranean. Ruffian or scholar, no medieval citizen would have known what a *non-agouti* gene was, any more than the cat would, but black cats must have seemed unusual in those days, quite suitable as a witch familiar.

Todd found that today black cats are more numerous in Great Britain, where cat torture was not so widely practiced, than on the Continent and in North Africa, which by medieval times was Muslim and may have been better disposed toward cats in general. According to legend, Muhammad so loved the cat Muezza that when she fell asleep on his sleeve while he was sitting down he cut off the sleeve rather than disturb her when he had to move. When he returned, Muezza bowed her thanks, and Muhammad in turn passed his hand down the length of the cat's back three times, conferring upon her and all cats thereafter the ability to fall down and land on their feet, right side up.

Along with all the irrational reasons for not liking cats, people who lived in cities where refuse accumulated had another reason for doubting that cats were truly helpful animals: they didn't seem to live up to their fabled abilities as rat catchers. Caravans traveling along the Silk Road and ships sailing from the East had, by medieval times, brought many wonderful things to the West in addition to silk: spices, tea, gems, apples, and other fruits — in short, all the luxuries that wealth could buy. But they, and later the Crusaders, also brought from the East less welcome goods. Just as trade in North America brought us corn borers unintentionally, so too, trade with the East brought rats of species never before seen in Europe. And with rats came diseases, including bubonic plague.

The rats, which had adapted in their native lands to live in cities alongside and within the buildings of men, thrived in medieval European cities with their scattered refuse heaps. No mere shy nosers-about-the-granary animals were these but self-confident feeders upon whatever humans cast out. They were aggressive and could inflict a painful bite. Rats grow and mature sexually at a rate that depends upon their food supply. When it is good, they can grow to monstrous size and produce frequent, numerous litters. In those days of considerable superstition, little understanding of natural history or the spread of disease, and casual sanitation, rats must have seemed true beasts of horror. The story of the Pied Piper of Hamlin hints at the helplessness that a town's citizenry must have felt as the rats grew in number and in dimension and in sass.

I felt a touch of that horror myself not long ago. I have no built-in fear of rodents, and in fact I spent the entire third grade, as did every girl in my class, playing mouse-mommy to a troop of waltzing mice, small black-and-white-spotted mutants with an anomaly of the inner ear that makes them run in circles. I kept them in a box at home and watched them run rounds through a cardboard tube. I thought them pretty little things and worried about them because they could not stop their circling (except, briefly, to mate and give birth —

their mutation did not seem to cut into their fertility). I also had as a childhood pet a white rat, his eyes, pink from lack of pigment, permanently squinted against the light. Herbie was a mutant form of the Norway rat, which originally came from Asia, and I had liberated him from a local pharmaceutical research lab. When I was in school Herbie lived in a birdcage quietly enough, but when I came home he took up his place on my shoulder. There he would cuddle against my neck and with his pointed nose explore the enticing hole that was my ear. I loved him and was disconsolate when one day he escaped through an opening into a wall. I never saw him again.

However, all that early rodent familiarity and affection didn't stop me from being unnerved by rats not long ago. My husband and I had gone to a play in Washington, leaving the car parked in a little-used alley in back of the theater. At the end of the performance, late at night, as we walked back to the car, I saw, scuttling about in that dimly lighted alley, the biggest rats I had ever seen, foraging on spilled garbage. They didn't threaten us in any direct way, but they didn't run, either. They were insolent, fat, sleek, and huge. They owned the alley that night. Something ancient inside made me shudder at the sight of them, and I confess I was glad to get inside the car and close the door against them.

How much worse rats must have seemed to people in medieval towns! Their defense was mainly cats, which had the *reputation* of being rat catchers. Stories and legends had long told of cats' prowess in this regard, and even today we are brought up on nursery tales of cats catching rats. Alas, that prowess may be overrated. James E. Childs, who is a researcher in infectious diseases at Johns Hopkins University, has studied the predation of feral cats on rats in Baltimore. Rats, he explained, have caused more human deaths as carriers of disease than all the human wars fought in history. With the help of the city animal shelter he first charted the areas of the city with the most feral cats and he began counting the cats in those places at night. He found that where there were the most cats there were also rats in the greatest numbers. That is not too surprising, when you consider that those places were rich in garbage, which both cats and rats found to be a good source of food. But why did the cats, legendary enemies of rats, placidly eat alongside them, a sight Childs saw over and over again? In the course of one thousand hours of night observation he saw cats kill rats only five times, and checks of cat dens and droppings further proved that the cats ate very few rats. For comparison he studied feral cats and rats in the city's parklands, where the pickings were less rich, and he found a higher kill rate there. But he also discovered something else. Comparing parkland rats (including the wood rat — *Neotoma floridana* — a New World native) with those in urban areas, he found that both size and onset of sexual maturity depend not upon their species but upon the amount of feed available. Urban rats, with their splendid sources of food, grow bigger and faster, and no matter the species or habitat, rats are capable of reproducing as soon as they reach seven ounces. Not only that but bigger rats produce larger litters. The pieces of the puzzle fell into place when he measured the few rats the cats had killed: they were all smaller, younger rats. Those that reached adulthood were too big for the cats to kill. They were no longer prey but were merely other animals with which the cats were forced to peaceably share spilled macaroni and cheese or leftover scrambled eggs.

Twenty-five years ago, when I was living on my farm, I had a beautiful but neurotic and twitchy calico cat (neurotic because her mother lived with us, too, and the mother batted and hissed at her daughter every time she saw her, which was often). The calico was, as many Ozark farm cats are, an excellent killer of snakes. She kept the chicken coop free of black snakes, which like to eat eggs in the nest. I also saw her once kill a poisonous copperhead snake right outside the back door of the house. She was a "far-darter" like Hekate that day, springing on the snake from a distance and expertly biting through the spine exactly at the base of the skull before he could turn and strike. She was also a considerable ratter. I knew that not only because the chicken coop and barn were free of wood rats but because she often left nearly intact carcasses outside the back door. But they were small country rats. Perhaps part of my shuddery reaction that night in the alley behind the theater was the flash of mental measurement I made. Those rats were huge, equal to a cat in size.

Medieval cities, lacking weekly trash pickup and tight-closing metal trash bins, would have been a prime place for rats to grow very large. Anyone abroad at night in one of those cities might have seen the same sight that James Childs did: cats and rats companionably rummaging through refuse. The Japanese turned against cats when the paintings of them failed to scare rats away from the silkworm-growing houses. Medieval city dwellers may have concluded that even real cats were not the rat killers the stories had claimed.

The rat grew larger as it became a fellow traveler of man, but the cat shrank. *F. catus* was descended from *F. s. libyca* individuals that were smaller than the run of the wild population, and by medieval times they had become considerably smaller still. And even those wildcats and early domesticated cats that earned the reputation of granary protectors were catching the smaller rats — the sort my calico caught with such dispatch. Although larger cats might have been better for catching rats, humans liked household cats to be smaller, and we altered them genetically to suit our fancy. The rats grew bigger. The cats grew smaller. The amazing shrinking cat.

By modern times, the Founder Effect had established several quite clear breeds or races of cats distinguished not only by coat color but also by hair length (the long-haired Persians, for instance) or other physical attributes (the tailless Manx). All these had been created by natural genetic mutations, which humans then preserved through selective breeding. During the late nineteenth century, breeders began deliberately to create new races for show and novelty. They worked to bring out recessive genes for anomalies of coat quality, coat color, and body shape: cats with bent ears, cats with kinky tails, cats with curly coats, cats with blue fur. Through that sort of specialized breeding, we have today the Singapura, or drain cat, whose ancestors lived in Malaysian drains. The world's smallest cat, it weighs less than six pounds. We have the Ocicat, which has the markings of an ocelot and was created in 1964 by an American who, working with a Siamese and an Abyssinian-Siamese cross, was attempting to create a cat that looked like a leopard. Another American breeder developed the California Spangled, so special that it was featured in a Neiman Marcus Christmas catalogue. A mutant kitten born to a normal black-and-white cat in Canada in 1966 lacked fur entirely. This proved to be the significant parent of a whole line of big-eared cats that are completely hairless and have worried-looking faces. They are said to be susceptible to sunburn and to suffer from the cold.

Cats have retained much more biological independence, overall, than have corn or silkworms during their association with us. Cats chafe at enforced confinement and ownership. They go feral quite successfully, as the armies of strays in any city suggest. Even if the rats are too big, other food is available. There are strays in the country, too, but they tend to get picked off quickly by larger predators. Maybe their famous independence came about because cats met us halfway in the domestication process: they came to us for the mice we had about us and then allowed us to admire them and take them into our homes.

Would cats pass the little-green-men test? If mankind were saucered away and returned five thousand years from now, would cats be missing, along with corn and silkworms? Certainly some kinds would; it is hard to imagine that a hairless cat could survive. Albino cats, with their eyes squinted against the pain of light, could be expected to die out under selective pressure in a world harsher than the sofa cushions. But very possibly some cats from special strains would remain. Of course, they would have continued to evolve and would not be the same as any dear kitty we now know. But they wouldn't be *F. s. libyca* either. Evolution doesn't run backward. It is a tinkerer, working with what is. We have irrevocably imposed upon cats new genetic modifications, created a new tune to play on the old four-note theme.

As to those rats: archaeological evidence has turned up some native European rat bones from the Pliocene and Pleistocene, but the Greeks and Romans had no name for rats as distinct from mice. They called all furry, long-tailed scurriers "mice" (that is, *Mus*) and included under that name, according to written descriptions, true mice, shrews, and even something that sounds rather like prairie dogs, as well as rats. Both Herodotus and Josephus, in reporting a battle between Assyrians and Egyptians, tell a story of a great troop of "mice" that one night descended upon the Assyrians and gnawed through their quivers, bows, and shield handles, utterly ruining them. If there is any truth to the story, that would have been the work not of mice but of rats, who were after the salty residues and leather. But the single generalized name used by the ancients would indicate that the rats they knew were probably of the mild-mannered sort, similar to our comparatively inoffensive wood rats, genus *Neotoma*.

Linguistically, the first distinction between rats and mice in Europe was made around A.D. 1100, and sometime between 400 and 1100 the black rat, *Rattus rattus,* appeared there. This species is thought to have originated in the wilds of Central Asia or the Middle East and followed trade routes to port cities of the Mediterranean. There it grew large and sleek and prolific on human stores and refuse. Being the climbing sort, it took happily to life on ships in harbors and began traveling wherever the ships went. Once in Europe, black rats spread rapidly from city to city, to the consternation of the citizens, who, after a time, began to make the association between rats and plague. The legend of the Pied Piper is thought to have originated in the 1200s. Rat catchers, making up for what cats failed to do, became important municipal officers, and in some places special taxes were levied that could be paid only in rat tails.

Black rats continued to be a problem for hundreds of years until, in the eighteenth century, their only effective predator migrated into Europe. It, too, was from the East, probably Mongolia. It was bigger, heavier, and more aggressive than the black rat and could outbreed it. Its name is *Rattus norvegicus,* the brown or Norwegian rat, so called for the place where the specimen for naming was taken. The brown rats thinned the black rat population considerably but were not an improvement at all from the human standpoint. Brown rats have all the worst features of black rats and then some.

When the colonists came to this continent, rats of both species accompanied them, and today they have spread throughout the world. A census is impossible, but there are estimates by sane and sober experts that for every inhabitant of New York City — and there are seven and a half million of them — there are somewhere between one and six rats. That is a lot of rats.

The domestication of silkworms and many other animals was our doing. That of cats may have been a two-way affair. But rats, and a number of our other companions, are something else entirely. You could say, by stretching the meaning of the word only a little, that rats and those other animals have domesticated us. We often dwell on the harmful effects we have on the rest of the species with whom we share the planet. But many species have benefited enormously from human presence and activity. If I am wrong about our ability to solve the problems of limits, they will miss us when we are gone. They will stick with us right to whatever end we engineer for ourselves. In addition to rats, a sampling of these organisms includes cockroaches, crabgrass, crows, and coyotes, as well as sea gulls, pigeons, raccoons, and larder beetles. A time may come when they will *be* our natural history.

Of Apples in
Heaven's Mountains
and in
Cow Pastures

A bad woman can't make good applesauce.

— Ozark folk saying

O NE AUTUMN MORNING not long ago, I was walking down a row of apple trees espaliered to wires
near Geneva, New York. I was being led by Philip
L. Forsline, the USDA's curator of apples and sour cherries,
through what is called the Core Collection, the most agri-
culturally interesting varieties of the three thousand apple
trees growing at the Plant Genetics Resource Unit of the
Agriculture Research Service at Cornell University, the biggest living library of apple trees anywhere in the world. As we
walked we took bites out of apples growing on the trees.

The day was cold and the wind made it seem colder. The
sky was leaden, promising an early snow, but maples in full
autumn color ringed the field and echoed the cheerful reds,

OPPOSITE: Gathering apples in the fourteenth century. Adapted from an
illustration in *Tacuinum Sanitas* [table of health] *in Medicina*.

russets, and yellows of the apples. The trees had been pruned to stand neatly in narrow rows. Some had drooping limbs, some grew straight and stiff. The shapes of the leaves and the color of the bark varied, as did the fruit, some clustered, some dangling independently. Some apples were huge, some not even bite-sized. The trees and the fruit they bore looked so dissimilar from one another that it was hard to believe that the entire group was botanically related. Some didn't look at all like my mental image of what an apple tree should be.

Bite and chuck. Bite and chuck. My fingers were growing cold and stiff, and the apple names I was attempting to write down turned into illegible scrawls. Some apples tasted so good that I wondered why they weren't in markets, but their names were unfamiliar; I stuffed my pockets with them. Bite and chuck. Some were so sour or bad-tasting that I quickly understood the apple growers' category "spitters." But Phil told me that even bad-tasting ones could be of interest because of their manner of growth, time of bearing, hardiness, or resistance to disease and pests. They could become part of the stock to help produce apples that tasted good and also had other important qualities.

Commercial apples are a serious business in the United States, which is the second-largest apple-producing country in the world. The first is China, which produces twice as many tons as we do. Serious agribusiness requires serious research, and that is what is going on in this USDA experimental plot. Putting in a commercial orchard or replanting an old one with a new variety is very expensive in terms of money, time, and labor. It takes a number of years before the new trees bear sufficiently to pay back the orchardist for his investment, even if the market is good, so apple developers need to be sure of the genetic qualities being packed into a new variety of apple before they promote it.

Phil, tall and rangy, with a sincere midwestern face, handed me yet another apple. "Now this," he said, "is the Lady apple. It is an ancient clonal stock from Roman times — probably brought to Rome along the Silk Road." Evidence suggests that apples originated in the mountains of Central Asia and were yet another trade good that spread both east and west along that highway. Phil was striding on ahead of me, talking apples, but I was lost to him for a moment, contemplating the experience of holding in my hand and eating the reincarnation of an apple eaten for dessert by a Roman at a banquet.

Today, when we use the word "clone," the mind wants to skip to a sheep named Dolly. But the Romans, who may have invented grafting (other candidates for credit include the Greeks, Persians, and Chinese), did cloning, and so have orchardists ever since. Grafting involves taking a shoot (orchardists call it a scion) from a tree that bears good apples and binding it to, or inserting it into a cleft in, the trunk of a tree that doesn't produce good apples but that has other desirable properties: vigor, resistance to disease, or the ability to dwarf and stay manageable, say. The new grafted shoot will grow up to produce the same good apples as the tree it was taken from. It is a clone of that tree; it is simply growing on another root system.

Grafting is an ingenious way humankind discovered to make an end run around apple genetics, which is, from our standpoint, perversely complicated. So complicated, in fact, that nearly every apple tree that grows up from a seed is a new variety, one whose fruit is not at all like that of the tree from which the seeds came. There are other ways to get around the problems of apple genes. For instance, whole trees can be grown from parts of the old one rather than from seeds. That is another kind of cloning. And, in fact, a few cultivated apples don't even have seeds.

The fruit of a tree — the part we call an apple — is the vegetative growth of the mother tree, the protective layer it packs around its seed. So the apple that we eat comes from the maternal line, while the seeds, and only the seeds, are the result of the sexual pairing of two parent trees. Some plants are self-fertile, which means that the pollen they produce can fertilize their own flowers. But with rare exceptions (Phil had pointed out one apple tree that was self-fertile) apples are self-*in*fertile. They need pollen from an entirely different tree to set fruit. When wind and bees are carrying pollen from one tree to another, the orchardist has no idea what qualities the male tree is writing into those seeds. The pollen may be from one of those small, sour apples popularly called "crab" (sweet apples and crab apples, of which there are also many species, cross so happily that their separate species names seem suspect). If so, its seeds might grow into a tree that would produce tiny, sour fruits. Or it might not. The orchardist would not have even a hint of the fruit that would come from a tree grown from seed that was openly and freely pollinated. Open pollination can be avoided by enclosing a tree and hand-pollinating each flower with a designated pollen, but even then there are surprises. So orchardists, who need to know dependably what sort of apples they will have for customers, put in proven grafted stock and never, never,

never save apple seeds to plant unless they have a speculative, adventuresome, experimental set of mind. But there isn't much room for experiment in commerce; orchardists don't like surprises.

I'd missed some apple lore by the time I caught up with Phil, but when I did he showed me a tree he called the Oregon crab apple. It grows, he said, on the West Coast all the way up to Alaska. He thought it might have been brought to this continent over the land bridge from Siberia long ago, when the sea level was lower, because it has some features that are similar to the apple I'd come to see, an apple that grows wild in Kazakhstan, conveniently near the route of the Silk Road. An apple with no common name, it is known only as *Malus sieversii*. Many researchers, including Phil, believe *M. sieversii* is the principal ancestor of *Malus* × *domestica,* which is the name for all the varieties of apples we buy at the supermarket. All cultivated apples come from trees that are genetic crosses, which are then grafted upon yet another species of root stock. They are, as some apple people point out, the products of "artificial trees."

Malus × *domestica* is the kind of name that makes some systematists shudder, protesting that it violates the rules of taxonomy because the × stands for a cross between, or hybrid of, at least two "true" species. A rigorous taxonomy does not recognize crosses or hybrids. But such rigor need not concern us here because an × in a species name is common among horticulturists and agronomists. In the case of apples, the parents of the crosses often were the kinds of trees that grew up in fencerows from seeds planted by birds in their droppings, the trees that grew in old pastures after the seeds had passed through a cow's gut. They were the trees that grew up out of the mash left over from cider making or from the apple core chucked aside by someone walking down a country lane, or even from seeds handed out by Johnny Appleseed. These trees often are what we call wild apples, and the fruit of most of them is sour and bitter (Thoreau claimed that wild apples taste good only when eaten while walking outdoors on a cold day). But occasionally one of those accidental trees produces a fine apple, a marketable one. Red Delicious, Golden Delicious, and McIntosh apples first grew on trees discovered quite by accident growing on old farms. And it is those three trees, along with a very few others, that agronomists have worked with to create all the apples we find in the supermarket.

Apples are so complex genetically that in the past, bewildered botanists suggested that they did not obey the rules of Mendelian genetics, and it is true that the progeny grown from the seeds of any apple tree do not sort out into the neat, predictable pattern that Mendel first laid out with his pea plants. We now know quite a lot more than Mendel did about "factors of inheritance," as he called genes. He was lucky. With his pea plants, he chose single traits (flower color, for instance) that are controlled by single genes. But a single trait can be affected by more than one gene (which is part of the reason that inheritance had been assumed to be a blending of factors); a whole series of genes turning on and off enzymatic processes may determine the genetic expression of a trait or biological function. And a given trait can be modified by other genes, as is the case with human eye color. Most of us have either brown or blue eyes, the result of a particular gene with two alleles that determine the amount of the brown-black pigment melanin in the front cells of the iris. Brown-eyed people, who have the dominant allele of the gene, have lots of melanin in the iris. Blue-eyed people, who have two recessive alleles, lack the melanin; their eyes look blue because the black pigment at the back of the eye shimmers through the clear iris as blue. But we all know people with eyes that are neither brown nor blue but hazel, gray, black, or some other mixed color. These are not the result of other alleles of the basic eye-color gene but of a whole complex of modifier genes, some of which control the amount of melanin that is laid down in the iris. Others affect the way the pigment is arranged, whether in clumps or uniformly. In fact most traits and not just eye color are affected by the interplay of a number of genes and are not simply the expression of a single one.

In addition, certain genes are expressed only when environmental conditions activate them — that temperature-related darkening of Siamese kittens' paws, for instance. Some traits appear in offspring not as independently and predictably as the colors of pea blossoms, but assort through linkages and require sophisticated statistical analysis — of the kind Marian Goldsmith and her colleagues are using in silkworms. And that is just for starters. Apples are even more complicated.

Back in my high school sex-education classes, we were taught the cozy rule of two. It takes a pair of human beings to make a baby, and humans have cells with twenty-three *pairs* of chromosomes carrying corresponding genes on each pair. That struck us as right and proper and consistent: two for the big picture, two for the little picture. But, as we youngsters were just beginning to learn, in most matters there are exceptions. In this case, the exception was that the cells for reproduction, our very own eggs and sperms, did *not* have paired chromosomes but single ones. In those confusing but lubricious teenage years we thought of the single chromosomes as lonely and seeking to come together to combine with their opposites — girls and boys, eggs and sperms, making pairs. When egg and sperm did pair and merge, a part of the father's attributes, packaged in each of his single chromosomes, and part of the mother's, similarly packaged in each of her single chromosomes, came together. They became a new cell, which now had paired chromosomes. Hey, presto! The comforting rule of two was restored. The fertilized egg, having the proper number of chromosomes, began dividing. And, though this account is colored by a vague adolescent lust that did not yet know the word "anthropomorphism," it is reasonably accurate as far as it goes. Many forms of life follow the same companionable rule of two, but others do not.

Among those that do not are the honeybees whose inheritance puzzled Mendel. The single egg-laying female bee in the hive, the queen, does have paired, or diploid, chromosomes. But the male bee, the drone, does not; he has unpaired single, or haploid, chromosomes. Most types of apples are diploid. In more cases than not, apples have seventeen pairs of chromosomes, for a total of thirty-four. But some varieties are haploid, with seventeen single chromosomes. Some, especially among the crab apples, are polyploid, which means that their chromosomes are not paired but tripled, quadrupled, or even wadded up into sixes. Some apples have up to eighty-five chromosomes. And each of the genes on each of the chromosomes can have different alleles. The result is that a whole bunch of genetic variation can be packed into a single seed, variation that has accumulated down through the ancestral lines. Such a rich genetic heritage puts unpredictability into every seed.

They didn't tell us about polyploidy in sex-education classes because we didn't need to know about it for human reproduction, and, in general, most other animals are diploid, too. As always, however, there are exceptions: for example, those haploid male honeybees as well as triploid lizards. And polyploidy has been induced artificially with silkworms and other experimental animals.

Among plants, polyploidy is not at all uncommon, and when it occurs, the results of plant breeding can be unpredictable. Polyploidy can be induced artificially, usually with chemicals, to produce what are called chimeras. New kinds of crop plants can be developed through the creation of chimeras. Ordinary diploid alfalfa, for instance, has a small leaflet, but tetraploid alfalfa, the one usually grown as hay for livestock, has bigger leaflets and can withstand drought and other stresses better than the diploid variety. Octoploid alfalfa produces a still bigger leaflet, which would seem to be a farmer's dream, but the plant is so tender it cannot stand field conditions and will grow only in a greenhouse.

Plant-evolution biologists suspect that spontaneously occurring polyploidy has produced new species in the past. Plants are generous and tolerant and easygoing about accepting extra chromosomes when different sorts pair, even if later on down the line they may rearrange them a little. Spontaneous polyploidy or some other shuffle of chromosome numbers may have been what made apples in the first instance. Apples belong to the Rosaceae, a large and diverse family of more than two thousand species of plants, including not only roses and apples but spiraea, strawberries, pears, and the stone fruits: plums, cherries, and apricots. Within that family, apples have an unusually large number of chromosomes, that typical thirty-four in seventeen pairs. So it has been suggested that one of the stone fruits (with its eight pairs of chromosomes) might have hybridized with one of the spiraeas (nine pairs) to produce the seventeen pairs of the ur-apple.

Even though the ur-apple may have rearranged its extra chromosomes and settled down to the usual diploidy in that practical, unfussy way in which plants go about life's business, many familiar varieties of apples that we buy in the market are polyploid, some of them spontaneously so: Stayman, Jonagold, Baldwin, and the beloved old pie apple Rhode Island Greening. The Jonagold is a modern, contrived cross between Jonathan and Golden Delicious parents. But the Stayman sprang up quite by accident, all on its own, from a Winesap seedling in Kansas, where it was discovered by a physician and plant breeder, J. Stayman, in 1866. The Baldwin and the Rhode Island Greening were spontaneous polyploids, too.

Because commercial varieties of apples today are the progeny of just a few varieties — a dozen at best — there has been a considerable restriction of the genetic base available to agronomists, compared to the 1800s, when authorities estimate that more than seven thousand cultivated kinds of apples were being grown and grafted.

"During the last thirty years, breeding objectives [for apples] have focused on meeting aesthetic standards established by supermarkets," wrote Dominique A. M. Noiton and Peter A. Alspach, horticulturists and agronomists. Orchardists with heavy investments in the standard trees lament that the market for apples is shrinking because the modern commercial varieties no longer have much flavor or pizzazz. Delicious apples are not delicious but supposedly present well, although an experience I had recently makes me doubt even that.

I was waiting to meet a friend in the lobby of a fine old hotel. I was sitting near a highly polished table set off by a basket of shiny Red Delicious apples for anyone to take. During the course of the quarter-hour I was there, a number of people walked by. Some stopped, not to take an apple but to look at them, commenting on their waxy, perfect, uniform appearance: "Aren't they beautiful!" But the more usual comment was a question, directed vaguely at me because I was sitting next to them: "Are they real?" And when I replied that I believed that they were and were free for the taking, the reply was some variation on "You know, they might as well *not* be real for the way they taste."

This is the sort of statement that depresses orchardists, particularly those in Washington State who have spent large sums for advertising and public relations over the years to convince people that Red Delicious is synonymous with "apple." It is also why in recent years so many horticulturists and others have devoted themselves to tracking down old varieties of apples on abandoned farms and in old cow pastures, taking seeds and cuttings to expand the genetic base from which apples can be grown. And it is why the USDA is sending scientists to Kazakhstan, the presumed birthplace of eating apples, to collect what is there before the wild apple forests disappear.

That was how I came to be in Phil Forsline's company that cold October day. I wanted to see and taste the wild apples that have been grown in Geneva, New York, since 1989, when the USDA first began collecting them in Kazakhstan. Phil has been one of the chief collectors.

After walking through the Core Collection, we got back in the car and warmed our hands as we drove over to the plot where the *M. sieversii* trees grow. The trees from the first collecting expedition outside the Kazakhstani industrial town of Almaty (formerly Alma Ata, one of the ancient way stations on the Silk Road★) are mature enough now to produce apples. In Geneva the trees are grown in tight, close rows, which keeps them more compact and uniform than they are in their native forests. But allowing for that, even to an untrained eye the individual trees showed obvious differences. The leaves were of various shapes and hues, the trees branched and twigged in different ways. Some sent up many stems, giving the trees a shrubby-looking appearance. Phil slit

---

★ Alma Ata means "the Father of Apples." The modern city, built in 1854 on the site of the ancient Silk Road town, was named Vyerni, and it kept that name until 1927, when it reverted to Alma Ata, after the ancient trading station. As linguistic archaeology, however, the name does not add up to a great deal. King Arthur's Avalon also means "Homeplace of Apples" in Welsh, and a brief flip through my desk atlas reveals that there are at least five Appletons in the United States, where nothing but crab apples grew until Europeans started bringing eating apples here as they settled the continent.

open a twig with his fingernail and showed me a red-streaked interior unlike the others. He then showed me a nearby plot of new trees. They were too young yet to bear fruit — so young that they were thorny, betraying their relationship to roses. Phil is particularly eager to see how these young ones turn out. "The oldest trees, the ones collected first, were taken near Almaty," he said pointing toward the big trees we had first looked at, "and I suspect there may be some introgression [hybridization with cultivated trees] among them, but these young ones are from seeds I collected a long way from Almaty, in places that have never been farmed."

Some of the new seedlings already were showing resistance to various apple pests and diseases, Phil told me, and he had seen that while he was collecting, too. "The interesting thing was that since the trees in the apple forests grow up from seeds, each tree is different from its neighbor. And you can see a healthy tree growing right next to a diseased one. Resistance is genetic. We think it would be good if we can incorporate it into commercial varieties of apple."

Some of the older apple trees from Kazakhstan, the ones growing in tight rows, bore fruit — apples of different shapes, colors, and sizes. Some were nearly as big as those you see in the store, some were as small as crab apples. When I was growing up, the certainty of a proposition was often expressed as being "as sure as God made little green apples." If the apples of Kazakhstan are indeed the offspring of the first and original ur-apple, God, or perhaps some other agent, made red, yellow, russet, purply red, and splotchy in addition

Malus sieversii growing in Kazakhstan.
Note the shrubby habit of growth.

to green ones. On his most recent trip to Kazakhstan, Phil had even seen white apples — apples with such a transparent skin that the creamy white flesh beneath was visible.

The wild apple forests of Central Asia are at high altitudes. As a result, the trees are very hardy and do well in a short growing season. On that October day in Geneva, most of the fruit on the mature *M. sieversii* trees was, in fact, past its prime, Phil cautioned as he handed me a medium-sized red apple to sample. It was no longer crisp and had a bland flavor, but still it was better than a Red Delicious.

When biologists try to determine where any plant or animal originated, they look for the place where it shows the greatest genetic diversity. You could call this the Founder Effect in reverse. Chinese silkworms feature a greater range of alleles than do those from isolated populations in other places, and this helped establish China as their place of origin. If you can trace the genetic range within an isolated population back to a population that apparently contains all the diversity found in populations everywhere else, it is a pretty good supposition that you've found the grandmama of all the populations, whether of silkworms or, in this case, apples.

Phil calls himself a horticulturist, not a geneticist, but he knows his apples, and in his travels collecting, from Almaty up into the Tien Shan range (the name means Mountains of Heaven), he has seen apples and apple trees that look like the whole range of those that grow elsewhere. They grow in rugged and inhospitable terrain where people have never farmed and few even have lived. As he described the wild apple forests, the area sounded like a giant agricultural demonstration plot that grew without human help. "You just stand there in one of those apple forests in the mountains where no one has ever been and you just *know* that this is where apples come from," said Phil.

In modern times apples of the species *M. sieversii* were first described by P. S. Pallas, a German naturalist who in 1786 saw them growing in the Caucasus, where, he noted, apple trees competed for dominance with oaks. It was Frank Meyer, an American who traveled the world looking for unusual plants, who in 1911 first took note of the apple forests in the Tien Shan. He had crossed into those mountains from China in the early spring and spent the next few months exploring the range, sometimes working at altitudes above ten thousand feet. He was chilled to the bone most of the time, but he kept on collecting, collecting, collecting. He was struck by the unusual forests of apples he found there, but no one followed up on his discovery until the 1920s, when Russian agronomists started investigating the area. Nikolai Ivanovich Vavilov, a geneticist who visited Alma Ata in 1929, was impressed with the wild apples he found growing near the town in "thickets," and observed, "We could see with our own eyes that here we were in a remarkable center of origin of apples." Being a geneticist in Stalin's time was a dangerous occupation, and because he stubbornly refused to transform himself into a Lysenkoist, Vavilov was arrested and died in prison.

It was another Russian scientist, Aimak Djangaliev, based in Kazakhstan, who initiated the cooperative exploration of those apple forests with the USDA and helped arrange the American scientists' trips there.

Recent molecular studies of *M. sieversii* have backed up the gut feelings of people like Phil Forsline and Nikolai Vavilov, showing that the species not only has more alleles of tested genes than other apples but also has a wider range of alleles at each gene locus.

Unfortunately, the apple forests of Kazakhstan are in danger of disappearing. "During the Soviet period," Phil said, "the area was preserved as national parkland, but with the breakup of the Soviet Union the mountains are no longer protected. Wealthy people are having many remote areas bulldozed and cleared in order to build vacation houses. The apple forests are disappearing, and the groves nearest to Almaty are 90 percent gone."

Those who see the apple's origins in Kazakhstan theorize that *M. sieversii* probably hybridized rapidly with crab apples of several species in the *Malus* genus that are native to Central Asia. It is those hybrids, not pure *M. sieversii,* that became the ancestors of what we now think of as proper eating apples.

But it is just those crab apples, tiny and sour, that some people see as evidence for other places of origin for the apple. Crab apples, despite their disappointing fruit, are very like eating apples in other respects, and they might, through genetic change, have come to produce apples that were sweeter and bigger — ancestors of domestic apples. They are native not only to Asia but to northern latitudes all around the world. There is tantalizing evidence of apples strung and dried, presumably sweet enough to eat, in Europe as far back as 6000 B.C., long before trade goods were making their way routinely from Central Asia.

Crab apples were cultivated in England for cider making long before the coming of the Romans, who are usually credited with bringing eating apples there. Some think that palatable eating apples might have sprung from those cultivated crabs in ancient days. A pre-Roman Saxon benediction asks that "the land may be filled with apples," although it doesn't specify whether they should be sweet or sour. There is a hoary tradition in Britain called Yuling (or Howling), a winter revel that took place after cider was made and suitably hardened. Men would rush out into the apple orchards with noise and jollity (and later with the shooting of firearms) to give thanks for the apple harvest just past and to chase away the wind from future crops. The custom survived well into Christian times, when it was reported that men would

run around and around the best-bearing apple trees with a
pitcher full of hardened cider, toasting them, and chanting:

> Stand fast root; bear well top;
> God send us a youling sop,
> Every twig apple big
> Every bough apple enow.

The question of whether apples began in one place or several may be settled before too long by a group of scientists at
Oxford University, who are analyzing the chloroplast and
nuclear DNA sequences of cultivated apples in order to un-
tangle their origins.

By the middle of the third millennium B.C. — about the
time the silkworm cocoon dropped into Empress Hsi-Ling
Chi's cup of tea and cats were at long last being invited into
Egyptian houses — tasty eating apples were being cultivated
far to the west of the Tien Shan range, as far even as Persia,
according to the archaeological evidence. Some centuries
later, apples had made their way to Greece and Turkey.
Whether dried or fresh, apples would have been an exotic
and hence desirable commodity along the Silk Road, as well
as a handy food for nomadic traders. Just as we do today, they
would eat and chuck the cores, spreading seeds that might
grow up into new trees along the way. *M. sieversii* also has a
botanical characteristic that allowed the planting of an orchard that would produce apples of the same quality as a
good mother tree, even without grafting onto proper rootstocks. The species is shrubby, sending up shoots from its
roots, and those shoots, when separated from the mother tree,
will root and grow into new trees. So will suckers from other

parts of the tree. If the shoots were carefully packed and kept moist, it may have been possible to establish orchards in Georgia, Armenia, and northern Mesopotamia from cuttings brought by merchant traders serving as de facto traveling nurserymen.

By the first millennium B.C., apples had become a standard part of the diet of the well-to-do in the Mediterranean world, and in the centuries following they became the dessert of choice at banquets, artistically heaped on the table in displays of many colors, shapes, and flavors. The Romans spread the knowledge of apples and their cultivation through the countries they conquered. By medieval times orchardists in Europe were so skilled that the privileged classes could offer fresh apples to guests every month of the year by growing some varieties that ripened early and others that were good keepers. The ownership of an apple orchard became a badge of prestige, as much a source of pride as a good wine cellar is today.

However, just as cats were both loved by the luxury-loving and tortured by others during the Middle Ages, apples, too, fell from favor among the less privileged. Popular wisdom, backed up by physicians, had it that apples caused "bad stomach" and fever, as well as "ill humors." This belief may well have been fostered by the apple's association with the fall of man in translations of the Bible, as in the famous King James version.

The writer of Genesis was a man of no particular botanical specificity, and in the oldest Hebrew and Greek texts, the Tree of Knowledge bears simply a generalized "fruit." That was the way it stood for centuries. In Latin texts of the early

The earliest illustration of the fall of man in which the
fruit offered by the snake is an apple. Adapted from the
depiction of a thirteenth-century French manuscript
in Miklos Faust, "Apple in Paradise."

Christian period, pictures of the expulsion of Adam and Eve
from Paradise show a generic-looking tree with tiny generic
fruits that look nothing like apples. The first known illustration that turns those fruits into apples is in a French manuscript from the thirteenth century. In it a snake with a
woman's head hands a round fruit to Eve, who in turn et cetera et cetera et cetera. The general suspicion of apples lasted
for several hundred years, and even after that they seem to
have carried a hint of the theological and moralistic. In 1603
Ralph Austin, an English Calvinist — and an orchardist —

put forth the notion that after the fall of man all apple grafts were loosened from their parent trees and what remained were crab apples.

By Austin's time the apple was back in favor and being grown and grafted in many varieties throughout Europe. The New World settlers, finding the place lacking in sweet apples, were quick to bring them to North America. Maine would like to have bragging rights to the first apple orchard in America. In 1604, just one year after Austin was writing, apple trees were planted by a Frenchman on St. Croix Island. The tiny island, today an International Historical Site, is slightly to the Maine side of the river of the same name, which separates New Brunswick from Maine and the United States from Canada. But New Brunswick could claim those apples as well.

At the time New England was being settled, orchards producing dessert-quality apples grew all over Europe, and those tending them had a sophisticated understanding of grafting. But the New England settlers grew apples not so much for eating as for animal feed and cider making,* and ungrafted apples can be used for those purposes. It may be that those first settlers, often urban sorts who had come as religious dissenters, were not skilled in rural crafts and were simply ignorant of grafting. At any rate, for whatever reason, most apple trees in America grew for a couple of centuries without ben-

---

* For us today, who take it for granted that potable water is available with the slight motion it takes to turn on the tap, it takes a mental effort to appreciate how important it was for our ancestors to be able to produce a beverage that would satisfy thirst without making them ill. Cider, fresh or hard, was such a drink.

efit of grafts. It wasn't until sometime in the 1800s that grafting became the regular way to grow an orchard.*

Those two hundred years of zestful seedling growth produced what John Bunker likes to call "the greatest genetic experiment ever performed by human beings." John may be forgiven some slight hyperbole because he is a dedicated collector of ancient New England apple varieties, particularly those of Maine, where he owns a nursery. Apple trees, in the right circumstances, can live two or even three hundred years, and John scrounges around the countryside for aging apple trees that have been preserved because they produce unusually good-tasting fruit. Many such trees may have grown from chance seedlings in the days of the early settlers. John and other collectors like him are interested in apples that are outside the criteria of the USDA, which does not add stock to its library of genetic resources unless the parentage of that stock is known. John collects seeds and shoots from trees of unknown ancestry.

"Apples in Maine," he told me, "came by two routes: from Massachusetts to the south and from Quebec to the north. Last summer I was collecting up near the Canadian border and found some very unusual apples there that had probably come from French varieties. One produced huge reddish fruit in which the calyx end [that is, the bottom] curves up to meet the stem so that when you slice it, it looks like a dough-

---

* For those interested in exceptions, it should be noted that Peter Stuyvesant, the last governor of New Amsterdam (New York City's predecessor in name), claimed to have grafted the first apple tree in America. The date was 1647, and the tree was on his farm, which today would be marked out by Fifteenth and Seventeenth Streets and First and Third Avenues. The tree was still bearing fruit when it was knocked down by a derailed train in 1866.

nut. It is a good drying apple and has a good flavor. No one knew its origins, but I was told that it had grown there for two hundred years. Another one I found in the north produced rectangular yellow apples about the size of a Red Delicious. Those apples had centers that were large and hollow. There was liquid in them which you could pour out as you ate the apple. I think the USDA people should look more in this country. I suspect that the genetic resources they are looking for are already present here."

John and others regard the period of free growth of seedling apples as a genetic experiment because the trees that grew up in cow pastures and fencerows and around old farmsteads were expressing the potential locked up in their genotypes inherited over the centuries. If they flourished, it meant that those genes produced hardy, thrifty trees able to withstand diseases and pests. And if they weren't cut down, it was often because the fruit they produced was tasty. That exuberant genetic experiment produced the Red Delicious, which was first noted in 1876 growing on a farm in Iowa. The Golden Delicious was discovered on a backwoods farm in West Virginia in 1890. Scrub growth on two different Massachusetts farms hid a couple of famous apple trees, which were discovered by accident: one was the Baldwin, found by John Ball in 1749; the other was noticed by Jonathan Hasbrouck, whose first name was given to the apple in 1826. The Winesap was found in 1817 growing up out of leftover cider mash in New Jersey.

All the centuries of growing grafted trees exclusively to produce apples with particular marketable characteristics, that end run around their genetics, has hidden and locked up a genetic potential that still has some interesting, as well as excellent-tasting, surprises. What else lies in the apple genome that we do not know about? We may have become too clever by half in narrowing the commercial genetic base.

Geneva, where the USDA's Plant Genetics Resource Unit is based, is a pretty town at the head of Lake Seneca. Once upon a time another repository of apple genetics, the kind John Bunker talks about, surrounded the lake, which is named for the Indians who lived there. The Seneca were famous farmers and orchardists, with homes in more than three dozen villages near the lake. They were part of the Iroquois Federation, which unfortunately sided with the losing nation in the Revolutionary War. General John Sullivan, a member of the First Continental Congress, was sent against the federation in 1779 and destroyed the Seneca villages. He was pleased to report that he had also destroyed sixty thousand bushels of corn, three thousand bushels of beans, and forty orchards of apples and peaches.

The apples that the Seneca had grown would have been from French varieties, for they received the seeds from Jesuit missionaries. It is not recorded whether the Seneca had learned how to graft, but it seems unlikely, because their trees had grown up as seedlings and few knew grafting in America at that time. Those who did, like Peter Stuyvesant, became famous for their skill. Another was William Blaxton, one of Boston's founding settlers. He was an eccentric clergyman and farmer who, despite an often-mentioned suspicious bookishness, became the best-known horticulturist in New England. He had arrived in 1625, set up his library, and then gone immediately to planting. On what is now the corner of

Beacon and Charles Streets in downtown Boston, he laid out
the colony's first apple orchard, which continued to bear
fruit for at least one hundred and fifty years. He was said to
have taken great delight in his bull, which he had broken to
saddle, and on that animal's broad back he enjoyed riding
about to visit friends and distributing apples and flowers to
them. Ten years after his arrival in Massachusetts, he moved
to Rhode Island and established its first apple orchard near
the present-day town of Pawtucket. There he grew an apple
that became famous in its time as Blaxton's Yellow Sweeting.
It is the presumed ancestor of the seedling that grew up in a
tavern yard at Green's End near Newport, owned by one Mr.
Green.

The apples the tavern-yard tree produced were well
known by the early 1700s as Rhode Island Greenings. Blaxton had passed on the knowledge of grafting to a few other
farmers, and the Greening was perpetuated through scions
taken from that tree in such great numbers that the tree was
said to have been killed by all the cuttings. The Greening
was still famous as a pie apple in my early days as a pie baker,
but I have not seen it for many years now.

By the late 1700s it had become a legal requirement for
settlers moving westward to plant apple trees, just as cultivating silkworms had been in some parts of the United States
even earlier. In order to receive a deed to his land five years
after settling on it, a newcomer to Ohio had to show that he
had planted fifty apple or pear trees. George Washington required prospective tenants on his land holdings to establish
one hundred apple trees. These orchards and others like
them, unlike those of Stuyvesant or Blaxton, grew without

benefit of grafting. John Bunker laughed when he told me the story of the McIntosh apple. John Mcintosh was a Massachusetts farmer who in 1811 emigrated to Canada, where he found on his new farm a tree that bore unusually good apples. He sold the apples and started planting its seeds so that he could sell young trees, too. Bunker wondered how McIntosh's customers felt, a few years later, when the seedlings began producing fruit that was nothing like the parent tree's. There must have been complaints, because in 1835 John McIntosh had learned how to propagate by grafting and was able to sell young trees that would yield the fruit he gave his name to.

Knowledge about grafting and the skill needed to do it gradually spread throughout the settled parts of the eastern United States during the 1800s. But to the west the great genetic experiment still had some years to run. Thanks for that go to the man we all learned about in bad poems in childhood: Johnny Appleseed.*

Growing up in Michigan, I remember people pointing out old apple trees growing in out-of-the-way places and saying that they had been planted from seeds given out by Johnny Appleseed. I've not seen anything to suggest that John Chapman ever made it to Michigan, but he did certainly pass through Pennsylvania, Ohio, Illinois, Indiana, and

---

* In these postmodern days, it is a cheap shot to quote Vachel Lindsay's "boom-a-lay, boom-a-lay, boom-a-lay boom" rhythms or, in this case, the tom-toms beating, beating, but his "In Praise of Johnny Appleseed" was an extremely popular poem in its time and was responsible for turning John Chapman into a folk hero. Lindsay did not have much early success with his poetry and for years tramped around the country exchanging poems for food. He must have felt Chapman was a kindred soul.

down to West Virginia, where tradition says one of his seeds grew into a tree that produced the Grimes Golden, an apple that was the foundation of the entire West Virginia apple industry.

Johnny Appleseed, who was born near Boston in 1774, became a convert, as did many people in his time, to Swedenborgianism, which had made its way to America by the 1780s. We need not be concerned with the finer points of the doctrine, which was named after its founder, Emanuel Swedenborg, who died two years before John Chapman was born. But the part that is germane here is that Swedenborg believed himself to be in direct, personal contact with God, who had revealed to him that there were two separate realities, one spiritual and the other material. The material — every stone or stick or animal, all the natural furniture of the planet — was caused by its counterpart in the spiritual world. This meant that every natural thing was an expression of the divine and should not be tampered with. It was just one of the theosophical ways of thought popular in that time, but to a greater or lesser degree many weighty figures of the 1800s called themselves Swedenborgians: Samuel Taylor Coleridge, Ralph Waldo Emerson, and the senior Henry James, for example. Later it touched the thinking and writing of poets Walt Whitman and Robert Frost.

John Chapman was more fervent than many others in his belief in the Church of the New Jerusalem, as organized Swedenborgianism was known. And so he gave up all comforts and set out on foot (barefoot, at that) to preach. He walked because he thought it would violate a horse's Divine Essence to ride it. He was, almost needless to say, a vegetarian.

He once doused a campfire because he saw that mosquitoes were being driven off by its smoke. When, in his travels, he saw farm animals being mistreated, he untied them and set them free. He preached from the Bible and Swedenborgian tracts and passed out apple seeds to all whom he met as a gesture of friendliness (and to gain their attention). Apples had a long history of use as a text for matters of the spirit. There was that association with the Garden of Eden, after all, in which the apple was not just an apple but a symbol. And Ralph Austin, the Puritan who wrote that on the occasion of mankind's first sin all apple grafts came asunder, had put forth that thought and many other spiritual ones about apples in a tract entitled *The Spiritual Uses of an Orchard.*

Seeds and Swedenborgianism made a perfectly good pairing for Chapman. But Johnny Appleseed seems to have truly cared about apples, because he also planted seeds along his way, leaving small orchards in his path, and it is estimated that when he died at the age of seventy-three, he had traveled more than one hundred thousand miles. Not all of those miles were on foot, though none were on horseback. Boats had no Divine Essence, so in 1806 Johnny Appleseed loaded onto one a supply of apple seeds from Pennsylvania cider mills and floated down the Ohio River to Wellsburg, West Virginia, where, with a brother, he established a nursery. From there he paddled the Muskingum River to the center of Ohio. In his travels he gave seeds to any settler and asked them to take care of the trees that grew up (Ohio had made their cultivation a legal requirement). And care, for Johnny Appleseed, was of the nurturing kind only. The Divine Essence of an apple tree, he firmly believed, was violated by pruning or grafting, so those settlers in what was then called the West were instructed to allow every apple tree to grow freely so as to express its spiritual reality.

In those years no one, let alone Johnny Appleseed, understood genetics. But whether or not those apple trees of West Virginia, Pennsylvania, Ohio, and points west expressed their spiritual reality, they did express the hidden reality of their genes. I suppose a Swedenborgian might say that the two were the same. At any rate, the great genetic experiment spread westward, and in the course of it the trees that were more susceptible to disease and pests died off. Most of the trees that did grow up produced small, sour apples, useful only for cider and animal feed. They were not too sour for cattle and wild animals to crunch down, however, and the seeds were spread by their droppings. And some of the trees turned out to produce perfectly wonderful apples, which went on to become the clonal parents of the apples we buy in the grocery store today.

By the time Johnny Appleseed died in 1845, grafting was becoming the preferred method of propagation for the serious commercial orchardist in this country, as it had long been in Europe. As growers gained botanical knowledge of the mechanics of apple reproduction, what came to be called "scientific breeding" began to produce valuable American strains of apples based on clonal varieties or, occasionally, on the rare find of an unusually good apple in a farmer's pasture. "Scientific breeding" narrowed the genetic range of apple growing and, in the interests of producing a marketable product, emphasized the appearance, shipping, and keeping qualities of apples rather than their flavor. And by limiting the genetic base, orchardists increasingly found themselves growing varieties that were susceptible to scab, mildew, brown rot, and a host of other diseases, not to mention codling moths, aphids, and spider mites, to name just three among the four hundred insects and mites that plague the modern grower. Those diseases and pests were fought with fungicides, insecticides, miticides, and a whole witches' brew of chemical sprays, dusts, and powders.

In the latter part of the twentieth century the public tolerance for solving agricultural problems with synthetic chemicals had begun to wear thin. In addition, for many crops, the diseases or insects against which the chemicals were aimed had been busily evolving resistance to it. Even when a pesticide was successful in fighting one pest, it sometimes brought about an even worse problem. In times past, red spider mites, for example, were never a big agricultural problem, but after insects that preyed upon them were killed (broad-spectrum sprays kill not only "bad" insects but "good" ones as well) they became a major pest for farmers. In addition, the chemicals that make modern, large-scale agriculture possible have become so expensive that in a tight commercial market they are a financial burden for a farmer or orchardist.

One of the reasons the USDA people are so interested in the apples from Kazakhstan is that many of them show genetic resistance to diseases and pests. Apple scab, to cite just one, is of large concern to orchardists. It is caused by a fungus that produces, quite literally, a scablike growth on the fruit. In its worst form it also makes the leaves fall off the apple trees. The apples do not sell well because customers have come to expect perfection in the appearance of food. In the past, scab was controlled by a fungicide, but because of the growing disenchantment with pesticides, considerable research is being done to give apples scab resistance directly as part of their genetic makeup. Giving an apple, or any other agricultural plant, resistance through crossbreeding isn't too difficult when the resistance is the responsibility of a single gene. Some crab apples have single-gene resistance to scab, and they can be crossed with eating apples. But, unfortunately, single-gene resistance does not last long in agriculture because crops, including those in orchards, are grown in such great numbers that the pest species evolves to overcome it. When resistance comes from several genes, that is, when it is polygenic, it lasts longer (nothing is forever: life evolves). But that is harder to achieve by crossbreeding. Single-gene crossbreeding can be done by making the original cross and then backcrossing to eliminate traits that may have come along with the resistance gene, perhaps smallness and sourness. But when the trait is polygenic, it may take five or six backcrosses to eliminate the other genes that have hitchhiked along with the ones helping to confer resistance and keep that variety's good qualities. An apple-tree generation is about four years (that's how long, on average, it takes to bear fruit), so the development of polygenic resistance would take something like twenty years as well as a plot of land big enough to grow and prove many crossbred trees.

That explains why young apple breeders are keen on what you or I would call genetic engineering. They don't call it that, however. They call it "transformation." Jules Janick, of Purdue University, and James N. Moore, from the University of Arkansas, wrote the standard text on fruit breeding. In the section on apples, Janick and some of his colleagues wrote, "To overcome these limitations of time and space, recent advances in biotechnology involving gene transformation now make it possible to introduce genes from almost any source into an apple." If the specific genes responsible for a certain trait can be precisely identified and directly transferred, the process is quick, efficient, and in some ways easier than crossbreeding. Transferring resistance directly from one kind of apple to another is difficult, but the possibility of using a resistance donor that is not an apple at all is looking more promising. A group of Belgian researchers put genes from radishes and onions into apples, choosing those donor plants because they had genes for the production of antimicrobial substances. It turned out that radishes worked best. And if a radish, why not a silkmoth? The cecropia silkmoth (not closely related to the *Bombyx mori* silkmoth) produces a peptide that attacks many kinds of bacteria. When the moth gene responsible for producing that protein was added to the DNA of apples used for rootstocks, the apples grafted onto them became resistant to fire blight, which is a devastating bacterial disease that turns apples, pears, and other plants black, as though they had been in a fire. And, at least on Gala and McIntosh apples, the same cecropia moth gene seems to help confer scab resistance, too.

The public did not like having the *Bt* gene put into corn. But the idea behind it was that *Bt* genetically produces a toxin harmful to moths and that adding that gene to corn would give the plant its own built-in resistance (at least for a time) to corn borers. Apple growers suffer from moths, too, and the same *Bt* gene is seen as a promising one to insert into apples to give the trees resistance to codling moths. These moths are a serious insect pest for orchardists for they turn maturing fruit into maggoty, unsaleable pulp.

I posed the little-green-men question to Phil Forsline. What apples would our returning descendants find on the planet when they returned after five thousand years of captivity by space aliens? Phil looked amused, reflected for a moment, and then said, "Well, they wouldn't find any of the eating apples that we've grown on this continent. Those need our care, and besides, they aren't native here. I don't think any of them would survive here." He paused, then added, "But they'd still be growing in Central Asia, and they'd be better."

In what way? I asked.

"The places where they grow now would no longer be threatened by development, and in five thousand years they'd have continued to evolve, growing hardier, more resistant to disease and pests, and better fitted to their particular surroundings."

He didn't say anything about the way their fruit would look or taste. That's the work of *Homo mutans*.

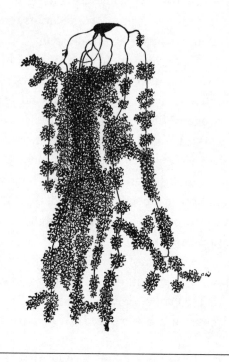

Afterword

Fellow traveler: translation of Russian *poputchik.*
One that sympathizes with and often furthers
the ideals and program of an organized
group . . . without membership in the group.
— *Webster's Third New International Dictionary*

HAWTHORNS ARE THORNY SHRUBS and trees that, like apples, belong to the big Rosaceae family. Although some are cultivated, most grow wild throughout temperate regions of the world. They are sometimes called thorn apples because their fruits look like brightly colored little apples. The picture-wing fruit fly, *Rhagoletis pomonella,* seems to have coevolved with the hawthorn, because it courts, mates, and deposits its eggs on hawthorn berries. But two species of parasitic wasp have also adapted to both the plant and the fly; they seek out hawthorn fruit containing the fly's eggs and, in turn, lay their own eggs

OPPOSITE: A nerve cell in a fly's brain. Adapted from a photo in Jonathan Weiner, *Time, Love, Memory.*

in the growing fly larvae. So efficient are the wasps that 90 percent of the fruit flies die as a result.

In the mid-1800s, land around the Hudson River Valley began to be cleared of wild growth, including the hawthorns. In the cleared land orchardists planted apple trees. The picture-wing fruit flies, already under pressure from the wasp predators, began to lose their hawthorn courting clubs and nurseries. That was serious for them, because the infant flies imprinted on hawthorn and searched for it later when they were ready to mate themselves. It is not physiologically impossible for them to reproduce on other fruit, but they are so tightly adapted to hawthorn that they grow and pupate only half as well on other kinds. Researchers have concluded that the flies' memory is triggered by the odor of hawthorns. Flies in general, and this species in particular, have big eyes and unusually good vision. Perhaps it was just a foolish fruit fly who first ventured to lay her eggs on an apple one warm end-of-summer day because it looked and smelled rather like a hawthorn. Perhaps it was a smart one with a sense of the taxonomic affinity of the two kinds of fruit. Or perhaps it was just an adventuresome one with a genetic hiccup in its hawthorn-imprinting mechanism. In any case, eventually new generations of picture-wing fruit flies began to emerge from apples and imprinted on apples as their host fruit rather than hawthorns. They didn't grow as well as the ones emerging from hawthorn fruit, but they had an advantage in that the parasitic wasps proved slower than they were on the adaptive uptake. To date, at any rate, the parasitic wasps do not bother the apples with the picture-wing fruit fly larvae in them. The apple-imprinted flies may not be as hearty but they are more numerous, and from the Hudson River Valley they have spread throughout the eastern United States. Apple growers don't call them picture-wing fruit flies, though; they call them apple maggots. When the female lays her eggs, she punctures the apples and the holes turn brown around the edges; the fruit inside becomes discolored and wormy.

The picture-wing fruit fly,
Rhagoletis pomonella.

It is not clear whether the apple maggot flies have yet become a completely separate new species, but they are considered at least a species-in-formation. One hundred and fifty years may seem like a short time for the creation of a species, but new species sometimes emerge even more quickly than that. These apple maggots are making their own Founder Effect, courtesy of us. They are yet another new species that we are creating with our fiddling.

In the old days, orchards were drenched with lead arsenate and DDT to kill apple maggots. Today other chemicals are used. But researchers are working on the introduction of genes into cultivated apples, in hopes of giving them resistance to the flies and other insects as well.

I find no record of any investigation into what has happened to the two species of parasitic wasps.

Malus × *domestica, Felis catus, Bombyx mori*. I chose these examples to tell the tale of how we have created new species because they are organisms that I've always been especially curious about. But I could have selected many others. The details would be different in each case, but the story would be similar. And with each species, as with the apple, the fly, the wasps, the lead arsenate and DDT, the orchardists, and the apple buyers, each story would involve unintended, unimagined, and even unknown consequences.

As I worked on this book, one of the scientists whose research I found enjoyable and instructive was Juliet Clutton-Brock, a British authority on the zoological aspects of domestication. I found her work on cats particularly useful and her writing insightful and graceful. In one of her papers she wrote, "My study of the history of human associations with wild and domestic animals has made clear that there are no wild places on the habitable earth, and almost no species of animal or plant, that have not been influenced by human activity. In the future, human interference with what is considered to be the 'natural world' is bound to increase."

I do not, of course, believe in the little green men. They will not come along and save the world from the effects of "human interference," with all its unintended consequences, problems of our own making. And even if they did, our descendants, when they returned to the planet, would commence altering and fiddling all over again. It is our nature. We are of this world, but we cannot live in it raw. Artifice is the way we get on. And we will continue to modify the genetic makeup of all kinds of life. We have been "genetic engineers" in the past, and we will continue to be so in the future. But the knowledge we have acquired since the elucidation of DNA and the laboratory tools we have created to use it now allow us to modify forms of life more precisely, more directly, and more quickly than in the past. Precision, directness, and quickness are what human beings are good at. What we have never been good at, in our past at least, is figuring out the impact, the consequences, of what our skills have allowed us to do. The century just completed has been a time of increasing technical skills, among them what we now call genetic engineering. Considering the needs of six billion human beings dedicated to ever-higher levels of consumption and extensions of life, genetic engineering will very likely have unprecedented unintended consequences. But that same century also witnessed the birth and the striking spread of ecological thinking. In science, ecological thinking *is* figuring out the consequences of one kind of life for others and for their environment.

This is an interesting and hopeful time in which to live, even more so to be born into. Our grandchildren are lucky. We are rapidly acquiring knowledge, if not yet understanding, of the genetic basis of life. Genes, it turns out, are simple. But the processes of life that they instruct do not yet seem to be. Until we can develop a deep, broad, and sensitive understanding of those processes, including an understanding of how forms of life are related to one another, we'll continue to suffer the unintended consequences of alterations. But if we do acquire that understanding and marry it to a broadened ecological intelligence, we'll have the mental equipment needed to solve the problems of limits.

I don't pretend to know the answers to those problems, but from looking at the past I do know that our relationship to other kinds of life has always been characterized, in varying proportions, by avarice, appetite, fear, and affection. Perhaps the answers will have something to do with increasing the proportion of the last item in that series.

APPENDIX

The following poem, by Bruce Haxton, proved too long to be a footnote in the text, so I include it here. I have never seen anything written so tellingly about the troubling, dark capacity we humans have to exploit other living beings and treat them cruelly simply because they are "other" and as such frightening in some deep way. But the poem is also optimistic, for it says that understanding can effect change. It is from Haxton's book *Dominion* (New York: Knopf, 1986).

What Would Make a Boy Think to Kill Bats?

I'd always thought I knew how bats caught bugs,
Big mouths opening spiked with rodent teeth,
But no, they scoop their prey up in midair
By cupping the web between hind legs and tail.
Watching them in slow motion made me think
How many nightfalls in the failing light,
How many nights in the succeeding shadow,
As a boy, I watched them browse and never saw
How for an instant with cupped wings and tail
They made their bodies into leather baskets,
Or how, in flight, they dipped their heads far in

To pluck the catch up out of the bottom,
Sometimes performing a somersault full tuck.

At twelve I found them — dozens draped heads down
Around the walls inside the dairy, sleeping.
Strange, I plucked them in their sleep with a BB gun.
Some had high-pitched barely audible screams
Which they made with jaws wide open. Others dropped
Without a sign. Why did I want them dead?

I remember the sight of them, how it was
Loathsome: lumps of dark flesh hanging from the wall,
Most of them rabid, I believed, things
Dangerous to have alive, and above all
Ugly, waking nightmares, although now,
When I remember the scene, as again
And again I do, the bats like warm-blooded angels
Unfold themselves with supple intricate wings,
With little cries of anguish, and no more
Can they frighten me, not now, not the bats.

SOURCES

If you are interested in reading more about any of the issues I've raised — and there is some really interesting stuff going on these days — any standard college biology textbook will help you get started. They are more fun to read than they used to be and are full of pictures and diagrams that help make everything clear. The one sitting on my desk is *Biological Science* by William T. Keeton and James L. Gould (Norton), but there are other good ones, too. For a historical perspective on what has been happening in genetics since the turn of the last century, I heartily recommend Jonathan Weiner's *Time, Love, Memory* (Knopf). It is a beautifully written book, ostensibly telling the story of the research life of one micro-biologist, but it will give you a broader understanding of the whole field. Richard Powers is a wonderful, thought-provoking novelist, and his *Gold Bug Variations* (Morrow) raises some of the moral questions associated with research on DNA. Below I list some of the more specific sources I have used for each chapter.

1. Of Humanity, Tazzie the Good Dog, and Corn

Corn Refiners Association. 2000. "Food and Industrial Corn Use — 1980 to Present." Internet printout.

———. 2000. "U.S. Corn Production." Internet printout.

———. 1999. *Corn Annual*. Washington, D.C.: Corn Refiners' Association.

Fussell, Betty. 1992. *The Story of Corn*. New York: Knopf.

Gray, Fred, et al. 1993. "U.S. Corn Sweetener Statistical Compendium." Washington, D.C.: U.S. Department of Agriculture, Economic Research Service.

Iltis, Hugh H. 1983. "From Teosinte to Maize: The Catastrophic Sexual Transmutation." *Science* 222.

Pacchioli, David. 1995. "Redrawing the Family Tree." *Research/ Penn State.* March.

Templeton, Alan P. 1989. "The Meaning of Species and Speciation." In *Speciation and Its Consequences,* ed. Daniel Otte and John A. Endler. Sunderland, Mass.: Sinauer Associates.

U.S. Department of Agriculture. Economic Research Service. 1999. "Sugar and Sweeteners." Washington, D.C.: USDA.

———. Forest Service. 1966. "Field and Laboratory Investigations of *Bacillus thuringiensis* as a Control Agent for Gypsy Moth." Upper Darby, Pa.: Northeastern Forest Service Experiment Station.

2. *Of Multicaulismania, Silkworms, and the World's First Superhighway*

Astaurov, B. L. 1972. "Experimental Model of the Origin of Bisexual Polyploid Species in Animals." In *Biologisches Zentralblatt* 91.

"Early History of Silk." 1938. *Ciba Review* 11. Basel, Switzerland.

Feltwell, John. 1991. *The Story of Silk.* New York: St. Martin's Press.

Forbes, Robert J. 1964. *Studies in Ancient Technology.* Vol. 4: *Textiles.* Leiden: Brill.

Franck, Irene M., and D. M. Brownstone. 1986. *The Silkroad.* New York: Oxford University Press.

Goldsmith, Marian R. 1995. "New Directions in the Molecular Genetics of the Silkworm." In *Recent Advances in Insect Biochemistry and Molecular Biology,* ed. E. Ohnishi et al. Nagoya, Japan: University of Nagoya Press.

Goldsmith, Marian R., and Jinrui Shi. 1994. "Molecular Map for the Silkworm." In *Silk Polymers,* ed. David Kaplan et al. American Chemical Society Symposium Series 544.

Goldsmith, Marian R., and A. S. Wilkins, eds. 1995. *Molecular Model*

Systems in the Lepidoptera. 1995. New York: Cambridge University Press.

Good, Irene. 1995. "On the Question of Silk in Pre-Han Eurasia." *Antiquity* 69.

Hewes, Gordon W. 1961. "The Ecumene as a Civilizational Multiplier System." *Kroeber Anthropological Society Papers* 25.

Howard, Leland O. 1904. "U.S. Department of Agriculture and Silk Culture." In *U.S.D.A. Yearbook, 1904.* Washington, D.C.: USDA.

Liu, Hsin-ju. 1998. *The Silk Road.* Washington, D.C.: American Historical Association.

Non-Mulberry Silks. 1979. United Nations Food and Agriculture Organization.

Procopius. 1979. *History of the Wars,* bks. III and IV. Trans. H. B. Dewing. Cambridge, Mass.: Harvard University Press.

Scott, Philippa. 1993. *The Book of Silk.* London: Thames and Hudson.

"Silk Moths." 1946. *Ciba Review* 53. Basel, Switzerland.

Strunnikov, Vladimir A. 1983. *Control of Silkworm Reproduction, Development, and Sex.* Moscow: Mir.

Tazima, Yataro. 1984. "Silkworm Moths." In *Evolution of Domesticated Animals,* ed. Ian L. Mason. London: Longmans.

———. 1968. "Evolution, Differentiation, and Breeding of the Silkworm: The Silk Road, Past and Present." In *Proceedings of the Twelfth International Congress of Genetics* 2, suppl. Tokyo.

U.S. Department of Agriculture. 1868. *Report of the Commissioner of Agriculture for the Year.* Washington, D.C.: Government Printing Office.

Wilson, Kax. 1979. *A History of Textiles.* Boulder, Colo.: Westview Press.

3. Of Lions, Cats, Shrinkage, and Rats

Alderton, David. 1995. *Cats.* London: Dorling Kindersley.

Armitage, P. L., and Juliet Clutton-Brock. 1981. "A Radiological and Histological Investigation into the Mummification of Cats from Ancient Egypt." *Journal of Archeological Science* 8.

Baldwin, James A. 1975. "Notes and Speculation on the Domestication of the Cat in Egypt." *Anthropos* 70.

Childs, James E. 1991. "And the Cat Shall Lie Down with the Rat." *Natural History,* June 1991.

———. 1986. "Size-Dependent Predation on Rats by House Cats in an Urban Setting." *Journal of Mammalogy* 67.

Clutton-Brock, Juliet. 1996. "Competitors, Companions, Status Symbols, or Pests." In *Carnivore Behavior, Ecology, and Evolution,* vol. 2, ed. John L. Gittleman.

———. 1993. *Cats Ancient and Modern.* Cambridge, Mass.: Harvard University Press.

———. 1992. "The Process of Domestication." *Mammalian Review* 22, no. 2.

———. 1981. *Domesticated Animals from Early Times.* Austin: University of Texas Press.

Eisenberg, John F. 1981. *The Mammalian Radiations.* Chicago: University of Chicago Press.

Hemmer, Helmut. 1990. *Domestication: The Decline of Environmental Appreciation.* Cambridge, Eng.: Cambridge University Press.

Hesiod. 1962. *Theogeny.* Trans. Richmond Lattimore. Ann Arbor: University of Michigan Press.

Hillaby, J. 1968. "Ancestors of the Tabby." *New Scientist* 38.

Hubbard, Alexandra, et al. 1992. "Is Survival of European Wildcats, *Felis silvestris,* in Britain Threatened by Interbreeding with Domestic Cats?" *Biological Conservation* 61.

Jerison, Harry J. 1973. *Evolution of the Brain and Intelligence.* New York: Academic Press.

Keeler, Clyde, et al. 1970. "Melanin, Adrenalin and the Legacy of Fear." *Journal of Heredity* 61.

———. 1942. "The Association of the Black (Non-agouti) Gene with Behavior in the Norway Rat." *Journal of Heredity* 33.

Kurten, Björn. 1971. *The Age of Mammals.* New York: Columbia University Press.

———. 1965. "The Carnivora of the Palestinian Caves." *Acta Zoologica Fennica* 107.

————. 1965. "On the Evolution of the European Wildcat." *Acta Zoologica Fennica* 111.

————. 1959. "Rates of Evolution in Fossil Animals." *Cold Spring Harbor Symposium on Quantitative Biology* 24.

Man and Animals. 1984. Philadelphia: University Museum, University of Pennsylvania Press.

Mery, Fernand. 1978. *The Cat.* New York: Grosset and Dunlap.

Metcalf, Christine. 1970. *Cats.* New York: Grosset and Dunlap.

O'Brien, S. J. 1980. "Extent and Character of Biochemical Genetic Variation in the Domestic Cat." *Journal of Heredity* 71.

Randi, Ettore, and Bernadino Ragni. 1991. "Genetic Variability and Biochemical Systematics of Domestic and Wild Cat Populations." *Journal of Mammalogy* 72.

Robinson, Roy. 1984. "Cat." In *Evolution of Domesticated Animals,* ed. Ian L. Mason. London: Longmans.

————. 1980. "Evolution of the Domestic Cat." *Carnivore Genetics Newsletter* 4. London.

Todd, Neil B. 1977. "Cats and Commerce." *Scientific American* 237, no. 5.

Turner, Dennis C., and Patrick Bateson, eds. 1988. *The Domestic Cat: The Biology of Its Behaviour.* Cambridge, Eng.: Cambridge University Press.

Van Valen, Leigh. 1970. "Evolution of Communities and Late Pleistocene Extinctions." In *Proceedings of the North American Paleontological Convention, Chicago, 1969,* ed. Ellis L. Yochelson. Lawrence, Kans.: Allen Press.

Williams, Robert W., et al. 1993. "Rapid Evolution of the Visual System." *Journal of Neuroscience* 13, no. 1.

Zeuner, Frederick E. 1963. *A History of Domesticated Animals.* New York: Harper.

4. Of Apples in Heaven's Mountains and in Cow Pastures

Brand, John. 1870. *Popular Antiquities of Great Britain.* London: J. R. Smith.

Browning, Frank. 1999. *Apples.* New York: Northpoint Press.

Cunningham, Isabel Shipley. 1984. *Frank N. Meyer: Plant Hunter in Asia*. Ames: Iowa State University Press.

De Bondt, An, et al. 1999. "Genetic Transformation of Apple for Increased Fungal Resistance." *Acta Horticulturae* no. 484.

Dickson, E. E., et al. 1991. "Isozymes in North American *Malus*." *Systematic Botany* 16.

Faust, Miklos. 1994. "The Apple in Paradise." *HortTechnology* 4, no. 4.

Forsline, Philip L. 1999. "Procedures for Collection, Conservation, Evaluation, and Documentation of *Malus* Germplasm." *Acta Horticulturae*, Proceedings of the 1988 Congress.

———. 1995. "Adding Diversity to the National Apple Germ Plasm Collection: Collecting Wild Apples in Kazakhstan." *New York Fruit Quarterly* 3, no. 4.

Hedrick, U. P. 1988. *History of Horticulture in America*. Portland, Ore.: Timber Press.

Hokanson, Stan C., et al. 1999. "*Ex situ* and *in situ* Conservation Strategies for Wild *Malus* Germplasm in Kazakhstan." *Acta Horticulturae* no. 484.

———. 1997. "Collecting and Managing Wild *Malus* Germplasm in Its Center of Diversity." *HortScience* 32, no. 2.

Janick, Jules, and James N. Moore. 1996. *Fruit Breeding,* vol. 1. New York: Wiley.

Johnson, George W. 1847. "The Apple." *Gardeners' Monthly* 11.

Juniper, B. E., et al. 1999. "The Origin of the Apple." *Acta Horticulturae* no. 484.

Lamboy, Warren F., et al. 1996. "Partitioning of Allozyme Diversity in Wild Populations of *Malus sieversii* L. and Implications for Germplasm Collection." *Journal of American Horticultural Science* 12, no. 6.

Lespinasse, Yves, et al. 1999. "Haploidy in Apple and Pear." *Acta Horticulturae* no. 484.

Noiton, Dominique A. M., and Peter A. Alspach. 1996. "Founding Clones, Inbreeding, Coancestry, and Status Number of Modern Apple Cultivars." *Journal of American Horticultural Science* 121.

Ponomarenko, V. V. 1986. *Data on the Knowledge of Apple Trees in the*

Caucasus. Trans. Geti Saad for the Agricultural Research Service. Karachi, Pakistan: Muhammad Ali Society.

————. 1983. "History of the Apple, Origin and Evolution." *Trudy po priklandoi botanike, genetike i selekstii* 76. Leningrad: Vsesoiuznyi institut rastenievodstva. (Summary in English.)

————. 1977. "The Specific Composition of Wild-Growing Apple Trees in the U.S.S.R. and the Centers of Their Genetic Diversity." *Botanicheskii Zhurnal* 62, no. 6. (Summary in English.)

————. 1974. "The Taxonomy of Some Species in the Genus *Malus.*" *Trudy po priklandoi botanike, genetike i selekstii* 52, no. 3. Leningrad: Vsesoiuznyi institut rastenievodstva. (Summary in English.)

Thoreau, Henry David. 1992. *Wild Apples.* Bedford, Mass.: Applewood Press.

Toussaint-Samat, Maguelonne. 1993. *A History of Food.* Trans. Anthea Bell. Cambridge, Mass.: Blackwell Reference.

Trifonova, Adelina S., and Atanass I. Atanssov. 1999. "Studies on Genetic Transformation of Apple." *Acta Horticulturae* no. 484.

Vavilov, Nikolai Ivanovich. 1997. *Five Continents.* Trans. Doris Love. Rome: International Plant Genetic Resources Institute.

————. 1951. *The Origin, Variation, Immunity and Breeding of Cultivated Plants.* Trans. K. Starr Chester. Waltham, Mass.: Chronica Botanica.

Watkins, Ray. 1976. "Apples and Pears." In *Evolution of Crop Plants,* ed. N. W. Simmonds. London: Longmans.

Way, R. D., et al. 1992. "Apples." In *Genetic Resources of Temperate Fruit and Nut Crops,* vol. 1, ed. James N. Moore and J. R. Ballington. Wageningen, Netherlands: International Society for Horticultural Science.

INDEX